Rice Biofortification

Pathways to Sustainability Series

This book series addresses core challenges around linking science and technology and environmental sustainability with poverty reduction and social justice. It is based on the work of the Social, Technological and Environmental Pathways to Sustainability (STEPS) Centre, a major investment of the UK Economic and Social Research Council (ESRC). The STEPS Centre brings together researchers at the Institute of Development Studies (IDS) and SPRU (Science and Technology Policy Research) at the University of Sussex with a set of partner institutions in Africa, Asia and Latin America.

Series Editors

Melissa Leach, Ian Scoones and Andy Stirling
STEPS Centre at the University of Sussex

Editorial Advisory Board

Steve Bass, Wiebe E. Bijker, Victor Galaz, Wenzel Geissler,
Katherine Homewood, Sheila Jasanoff, Colin McInnes,
Suman Sahai, Andrew Scott

Other titles include

Dynamic Sustainabilities
Technology, Environment, Social Justice
Melissa Leach, Ian Scoones and Andy Stirling

Avian Influenza
Science, Policy and Politics
Edited by Ian Scoones

Epidemics
Science, Governance and Social Justice
Edited by Sarah Dry and Melissa Leach

Rice Biofortification

Lessons for Global Science and Development

Sally Brooks

from Routledge

First published by Earthscan in the UK and USA in 2010

For a full list of publications please contact:
Earthscan
2 Park Square, Milton Park, Abingdon, Oxon OX14 4RN
711 Third Avenue, New York, NY 10017

Earthscan is an imprint of the Taylor & Francis Group, an informa business

Notices
Practitioners and researchers must always rely on their own experience and knowledge in evaluating and using any information, methods, compounds, or experiments described herein. In using such information or methods they should be mindful of their own safety and the safety of others, including parties for whom they have a professional responsibility.

Product or corporate names may be trademarks or registered trademarks,and are used only for identification and explanation without intent to infringe.

ISBN: 978-1-84971-099-2 hardback
ISBN: 978-1-84971-100-5 paperback

Typeset by FiSH Books, Enfield, UK
Cover design by Susanne Harris

A catalogue record for this book is available from the British Library

Library of Congress Cataloging-in-Publication Data

Brooks, Sally.
  Rice biofortification : lessons for global science and development / Sally Brooks. — 1st ed.
     p. cm.
  Includes bibliographical references and index.
  ISBN 978-1-84971-099-2 (hardback) — ISBN 978-1-84971-100-5 (pbk.) 1. Rice—Breeding.
2. Crop improvement. I. Title.
  SB191.R5B76 2010
  633.1'8233—dc22
                                    2010017151

To the memory of Donald Brooks
(1929–2005)

Contents

Abbreviations		*ix*
Acknowledgements		*xiii*

Introduction: Why Biofortification? 1
Global Science, Public Goods? 3
Biofortification as Biopolitics 4
Focus on Rice: Iconic Crop, Model Cereal 5
On Researching International Science Policy Processes 7
Chapter Preview 12

1 'Old Lessons and New Paradigms': Locating Biofortification 15
International Crop Research and the CGIAR 16
Pathways Linking Agriculture, Nutrition and Health 32
'Old Lessons and New Paradigms' 40

2 Building the Argument: The Case of Iron Rice 43
Introduction 43
A Win–Win Proposition: Nutrition and Yield 44
IR68144: 'A Serendipitous Discovery' 48
Iron Rice: The Silver Bullet? 51
Proof of Concept: The Sisters of Nutrition 53
From IR68144 to MS13: 'A Special Variety' 59
National Release: In the Shadow of Hybrid Rice 63
Conclusion 66

3 An Institutional Model? The Case of Golden Rice 68
Introduction 68
Rice Biotechnology: Laying the Foundations 69
Vitamin A Deficiency: Construction of a Public Health Problem 71
Golden Rice: A Scientific Breakthrough 74
A Science Policy Controversy 76
Granting Access, Keeping Control 83
Framing 'Acceptance': The Case of the Philippines 88
Conclusion 91

4   An Alliance Around an Idea: The Shifting Boundaries of HarvestPlus   93
    Introduction   93
    Back to Basics? A Challenge Program   94
    A Turning Point: Enrolling the Gates Foundation   100
    Establishing HarvestPlus   105
    HarvestPlus Comes to IRRI   107
    Interdisciplinary Encounters   109
    Brokers or Gatekeepers? Organizational Tensions
      and 'Global Science'   114
    Constructing Demand, Predicting Impact   116
    Impact and 'Spin-Offs'   119
    Business as Usual? The ProVitaMinRice Consortium   121
    Conclusion   123

5   Global Science, Public Goods? A Synthesis   125
    International Research Partnerships: Rhetoric and Reality   125
    Towards Interdisciplinary Integration?   128
    De-linking Impact and Context   130
    GM or Not GM – Is that the question?   133
    Boundary Terms and 'Escape Hatches'   135

Conclusion   138
    Locating and Engaging 'Users'   140
    Rethinking Upstream–Downstream Relations   141
    Towards a More Reflexive 'Public Goods' Science?   141

Notes   143
References   157
Index   171

Abbreviations

ADB	Asian Development Bank
AGM	Annual General Meeting
AGRA	Alliance for a Green Revolution in Africa
AIS	Agricultural Innovation System
AKIS	Agricultural Knowledge and Information System
ARI	Advanced Research Institution
BCP	Biotechnology Coalition of the Philippines
BMGF	Bill and Melinda Gates Foundation
Bt	*Bacillus thuringiensis*
CGIAR	Consultative Group on International Agricultural Research
CIAT	International Center for Tropical Agriculture
CIP	Centro Internacional de la Papa (International Potato Center)
CIMMYT	International Maize and Wheat Improvement Center
CP	Challenge Program
DA	Department of Agriculture
DALY	disability-adjusted life year
DFID	Department for International Development
DNA	deoxyribonucleic acid
DOST	Department of Science and Technology
ESDA	Epifanio de los Santos Avenue (main highway and ring road in Metro Manila)
ESRC	UK Economic and Social Research Council
ETH	Eidgenössische Technische Hochschule (Swiss Federal Institute of Technology)
FAO	Food and Agriculture Organization
FCND	Food Consumption and Nutrition Division

FNRI	Food and Nutrition Research Institute
FSR	farming systems research
GAIN	Global Alliance for Improved Nutrition
GBDS	Global Burden of Disease Study
GC9	(Gates Global) Grand Challenge No.9
GM	genetically modified
GR	Golden Rice
GxE	genotype by environment
G8	Group of Eight
HIV/AIDS	human immunodeficiency virus/acquired immunodeficiency syndrome
HKI	Helen Keller International
HRCP	Hybrid Rice Commercialization Program
HYV	high-yielding variety
IAEA	International Atomic Energy Authority
IARC	International Agricultural Research Centre
ICRW	International Centre for Research on Women
IDD	Iodine Deficiency Disorders
IDRC	International Development Research Centre
IDS	Institute of Development Studies
IFPRI	International Food Policy Research Institute
IHNF	Institute of Human Nutrition and Food
IFR	iron-fortified rice
IITA	International Institute of Tropical Agriculture
IP(R)	intellectual property (rights)
IPG	international public goods
IPRB	International Program on Rice Biotechnology
IR-	Designation given to all IRRI-developed plant breeding materials
IRRI	International Rice Research Institute
ISAAA	International Service for the Acquisition of Agri-biotech Applications
IVACG	International Vitamin A Consultative Group
MAP	Mexican Agricultural Program
MASIPAG	Magasaka at Siyentipiko Para sa Pag-unlad ng Akricultura (Farmer-Scientist Partnership for Development)

MDGs	Millennium Development Goals
MS-	Maligaya Special
MSG	monosodium glutamate
MTA	material transfer agreement
NARS (NARES)	National Agricultural Research (and Extension) System
NCT	National Cooperative Testing
NFA	National Food Authority
NGO	non-governmental organization
NIDs	national immunization days
NRM	national resource management
NSIC	National Seed Industry Council (formerly the Philippine Seed Board)
NSIC Rc- (PSB Rc-)	Standard designation given to rice varieties passed by the National Seed Industry Council of the Philippines (formerly the Philippine Seed Board)
NVASP	National Vitamin A Supplementation Program
OECD	Organisation for Economic Co-operation and Development
PATH	Program for Appropriate Technology in Health
PhilRice	Philippine Rice Research Institute
PRRM	Philippine Rural Reconstruction Movement
PSNL	(Federal) Plant, Soil and Nutrition Laboratory
PVMRC	ProVitaMinRice Consortium
QPM	Quality Protein Maize
RDA	recommended daily allowance
R&D	research and development
RTWG	Rice Technical Working Group
RVIG	Rice Varietal Improvement Group
SCN	(United Nations) Standing Committee on Nutrition
SEARICE	Southeast Asia Regional Initiatives for Community Empowerment
SUSTAIN	Sharing US Technology to Aid in the Improvement of Nutrition
SW/EP	system-wide/ecoregional programme
TAC	Technical Advisory Committee
TP	technical property
UN	United Nations

UNDP	United Nations Development Programme
UNICEF	United Nations International Children's Emergency Fund (also known as the United Nations Children's Fund)
UP	University of the Philippines
UPLB	University of the Philippines, Los Baños
USAID	United States Agency for International Development
VAD	vitamin A deficiency
VITAA	Vitamin A for Africa
WARDA	Africa Rice Centre (formerly West Africa Rice Development Association)
WHO	World Health Organization
ZEF	Zentrum für Entwicklungsforschung (Center for Development Research)

Acknowledgements

The research for this book was carried out between 2005 and 2008 and included around 90 semi-structured interviews with informants in the Philippines, China, the United States, Switzerland and the United Kingdom; participation in various meetings and conferences; and analysis of policy documents, press releases, biographical data, annual reports, scientific papers and published articles. It was supported by a grant from the Economic and Social Research Council (ESRC award no. PTA-031-2004-00017). I am grateful to the ESRC for this support.

I would like to thank the following people who all helped to make this research possible. Firstly I would like to thank all those who welcomed me into their homes when I travelled to the Philippines and the United States to conduct this research; especially Mavic and Martin Gummert, Bobbi and Yogi Thami, Margaret Hall and Bob Cole. I am grateful to Saturnino Borras Jr and Bel Angeles for introductions within the University of the Philippines and NGO community and Bita Avendaño and her colleagues for facilitating meetings within the International Rice Research Institute (IRRI). Special thanks go to representatives from International Food Policy Research Institute (IFPRI), IRRI and Zhejiang University who made it possible for me to attend the Second HarvestPlus-China meeting in Hangzhou in September 2007. Thanks also go to Lawrence Haddad for introductions to colleagues within the Consultative Group on International Agricultural Research (CGIAR) system and donor organizations and to Cathy Shutt for introducing me to her friends in the Philippines, connections which enabled me to participate in 'daily life' at IRRI in a way that would otherwise not have been possible.

I am also grateful to the many colleagues who have offered their intellectual support. First and foremost, I am indebted to Ian Scoones for his wise guidance and unfailing enthusiasm throughout the doctoral research that led to this book. I am grateful to Paul Richards for reviewing the manuscript. I would like to thank my colleagues in the Knowledge, Technology and Society Group at the Institute of Development Studies at the University of Sussex, particularly Melissa Leach, for their support and encouragement. Friends and family have also provided invaluable intellectual and moral support along the way; particularly Pauline von Hellerman, Jong-Woon Lee, Christina Oelgemöeller, Dominic Glover, Katarzyna Grabska, Ann Robins, Magda Tancau and David Thompson. I would especially like to thank my mother,

Pamela, and my sister, Pippa, for their continued support and tolerance of my frequent and often prolonged absences over the years; and my nephews, Duke and Joe, for our many hours of laughter, despite my having 'so much homework'.

Most of all I would like to thank all those people who shared their time, experience and insights with me in the course of this research. Particular thanks go to informants at the IRRI, Philippine Rice Research Institute (PhilRice), IFPRI and the University of the Philippines (Los Baños and Diliman), whom I met in many cases on more than one occasion, for being so generous with their time. I am grateful to the representatives from the many institutions I visited – universities, international development agencies, governmental and non-governmental organizations – for all I learned in my discussions with them. The responsibility of the content of the book is, however, solely mine.

Introduction:
Why Biofortification?

> Imagine a new breed of crops capable of alleviating malnutrition in even hard-to-reach rural populations – crops such as rice loaded with iron, wheat strengthened with zinc, and sweet potato packed with pro-vitamin A. These staples could be grown on family plots throughout the developing world. (HarvestPlus, 2004a, p1)

Biofortification – the enrichment of staple food crops with essential nutrients – has been heralded as a uniquely sustainable solution to the problem of micronutrient deficiency or 'hidden hunger'. Today, it is argued, poor people worldwide rely on staple crops such as rice, wheat and maize to meet most of their nutritional needs. By breeding or genetically engineering varieties of these crops with higher nutrient levels, the solution to this pressing global problem can be built into the seed itself and reach previously unreachable populations in the remotest areas of the developing world.

This is a story of competing pathways to better nutrition and health and how one vision, based on the promise of this 'new breed of crops', has begun to overshadow all others. This book traces the construction of a case for biofortification as a sustainable, cost-effective public health intervention, not by health institutions, but by members of an international agricultural research community. This point of departure has been highlighted through the framing of biofortification as a ground-breaking, boundary-crossing endeavour – an international, multi-institutional, interdisciplinary research initiative indicative of what has been called a 'new paradigm' (CIAT and IFPRI, 2002), in which agriculture is to be mobilized as 'an instrument for human health' (Graham, 2002).

While much is made of the new paradigm, the institutions at the centre of biofortification research are members of the international network of agricultural research centres that was the engine of the 'Green Revolution' of the 1960s and 1970s. Perhaps not surprisingly, there are echoes from the past. The notion that a solution to complex problems such as food shortage and rural poverty could be embedded 'in the seed' and therefore inherently scale-neutral was central to the Green Revolution, through which high-yielding varieties (HYV) of rice and wheat were disseminated throughout South and Southeast Asia and Latin America.

This was despite the reliance of these 'miracle seeds' on a package of chemical inputs and policy measures. Nevertheless, this popular narrative of the Green Revolution – the socio-technical transformation for and from which the Consultative Group on International Agricultural Research (CGIAR) became established as the international infrastructure for international 'public goods' crop research – has endured. Similarly, within biofortification research, pronouncements of a 'new paradigm' are accompanied by a logic that translates complex agriculture–health dynamics into a suite of nutrient deficiencies: deficiencies in humans and therefore, by extension, in the crops they plant and eat, and treatable as 'isolable problems' (Anderson et al, 1991) that can be solved by the application of the crop sciences, in particular plant genetics.

Focusing on initiatives in rice, 'the world's most important crop', this book explores the case of biofortification research as an exemplar of science-policy processes in international crop research. It highlights trends towards complex, heterogeneous research networks (in which CGIAR centres retain a central, if different, role) linking institutions, disciplines and sectors in new forms of partnership to solve problems of global significance through a return to 'upstream' research. As such, it raises pertinent questions for donors attracted by the absorption capacity of such networks, and in particular for 'new philanthropists' – or 'philanthrocapitalists' (Edwards, 2008) – attracted by the promise of a new generation of 'silver bullet' solutions that appear to offer 'impact at scale'.

This raises questions about the governance of science and technology for development, particularly in an era in which public-private partnerships are promoted as a way to compensate for the deficiencies of under funded public research systems. In this respect, this book offers insights into questions of technology governance that have tended to be crowded out of debates polarizing around the twin axes of private- versus public-sector ownership and control, and the use (or not) of transgenic techniques. In particular, it highlights the more subtle ways in which an international public system is transforming itself in preparation for more intensive engagement with the private sector, even as it reasserts its *raison d'être* in public goods research.

This analysis is particularly relevant at a time when concerns about declining agricultural productivity (particularly in Sub-Saharan Africa) and a global food crisis have raised the profile of debates about the role of science and technology in development. Today, after, some would say, years of neglect, agricultural research is back on international donor agendas. At the same time, a new generation of wealthy philanthropists, who made their fortunes in 'high-tech' industries, have turned their attention to the international development arena, bringing a new sense of hope to the search for science-based solutions to pressing global problems. While the spotlight is now on Africa, the starting point for these new initiatives is a familiar one; that the continent 'missed out' on earlier initiatives to stimulate agricultural production through investments in science and technology, from which populations in countries across Asia have already benefited, notably the Green Revolution. In this case, the contested nature of these prior 'successes'[1] – in socio-economic, political and ecological terms – has not featured in these debates in the drive to seize the moment and make the case for large-scale donor support.[2]

For some, this has provided an opportunity to repeat, with renewed urgency, their call for particular technologies to be 'embraced', or at least subjected to less rigorous systems of regulation, since they offer the best way – the only way – to increase productivity and ensure food security (Paarberg, 2009). This book argues that debates that focus on the relative merits of specific technologies, such as transgenic techniques or those of 'conventional plant breeding', miss the point, since they deflect attention away from more fundamental questions about an emerging model for global, 'public goods' science. If this model is allowed to become entrenched, it is likely to close down rather than open up future debates about the relative merits of competing technological pathways.

Global Science, Public Goods?

Biofortification research provides a lens through which to question the idea of 'global science', and the notion that it can generate generic research outputs as international public goods. Both notions have historically relied on an understanding of the innovation process as a linear path (Rogers, 2003), linking 'upstream' basic research with the later stages of adaptive research and adoption further 'downstream'. Within the CGIAR, these principles are articulated in terms of its 'comparative advantage' in conducting basic research, the outputs of which have sufficiently wide applicability that they can be disseminated as international public goods, amenable to adaptation and adoption in different parts of the world (Science Council, 2006). Nowhere have these assumptions been more explicit than in the design and justification of the Challenge Program model, for which the Biofortification Challenge Program, named 'HarvestPlus', is considered to have been the most successful pilot. The *raison d'être* for this programme model was to tackle enduring global problems requiring contributions from a range of disciplines and sectors. As such it provided a mode of organization for interdisciplinary integration, at the global level, that would ultimately deliver the required public goods.

These developments within the CGIAR can be seen as a response to broader trends in international development. Today, any initiative under the umbrella of 'development' must be justified in terms of a set of global development targets known as the Millennium Development Goals (MDGs). Initially the outcome of a contested process of weaving together 'different waves of earlier unachieved goals and promises made in various UN and other international summits and conferences' (Saith, 2006, p1169), this framework has, over time, been streamlined and 'black boxed' (Latour, 1987) as a global blueprint for development against which all prospective and existing projects are to be justified and their impact evaluated.

Concurrent with this shift towards goal-oriented development has been the envisioning of a new role for what might be called science for development. While presented to the public in terms of a simple formula of 'silver bullets' trained on development targets, global biofortification programmes are characterized by complex and still evolving organizational arrangements for delivering cutting-edge interdisciplinary science. Typically, these arrangements bring together different types of institutions – international and national, public and private, positioned 'upstream' and 'downstream' and representative of multiple disciplines, within global research 'partnerships'. In this context, the role of the CGIAR 'centre' is recast as that of 'broker' within these complex networks (Rijsberman, 2002, p3). Implicit in this redefinition is an assumption that, in taking on this redefined role, the CGIAR is uniquely placed to act as the honest broker, steering the direction of these networks in a direction that is consistent with its own public goods mandate.

Ultimately, biofortification research is justified in terms of its projected impact, understandings of which are increasingly framed by the MDG framework. Crucially, notions of upstream research as a generator of widely applicable, international public goods appear to satisfy MDG-framed understandings of impact in ways that have also found resonance within an acknowledged leader of a new generation of philanthropists, the Bill and Melinda Gates Foundation (BMGF) (Economist, 2006; Boulton and Lamont, 2007). Viewed from upstream, however, what does the term 'impact' mean to the actors, who are so far removed from the intended beneficiaries who define, predict and measure it? In this case, conventional linear innovation models provide a sense of – arguably misplaced – reassurance to actors, whose role it is to convince donors that all bases are covered and to convey a degree of certainty about impact, despite prevailing conditions that make such certainty yet more elusive. This book has followed a series of actors involved, in various ways, in biofortification research as they deal with the various uncertainties around the science, organization and impact of rice biofortification research on its, so far, largely imagined, beneficiary populations.

Biofortification as Biopolitics

The way in which the complex problem of micronutrient malnutrition has been repackaged as a 'challenge' to be put to a 'global' community of scientists highlights the importance of the way in which problems are framed, and in particular the inherently political process through which issues are understood and presented as 'technical' or otherwise. Framing is understood here as a 'core discursive activity' (Apthorpe, 1996, p24) which involves 'matters of inclusion, exclusion and attention, including how the burden of proof is distributed, and the perception of alternatives and constraints' (Gasper, 1996, p47). This perspective eschews conventional, linear science policy and innovation models, drawing attention to the way in which (often unstated) normative commitments shape the way problems, solutions and target groups are co-constructed at the outset, in such a way as to frame out a range of issues and options from subsequent debate.

In the case of international biofortification research, a complex, multi-dimensional social problem – malnutrition – has, through the reductionism of nutritional and agricultural sciences, been translated into a series of nutrient deficiencies. Specifically, biofortification locates the problem in the seed – now revealed as deficient in required nutrients – and the solution in a programme of plant breeding and genetic engineering of micronutrient-dense crops. This convergence of human and plant biological sciences on standardized, reductive approaches, encouraged by prevailing goal-based frameworks for the global management of nutrition and development, can be understood in terms of global 'biopolitics' (Foucault, 1976; see also Brooks, 2005). While existing large-scale micronutrient interventions such as food fortification and supplementation already follow a 'medical model' (Delisle, 2003), this vision goes a step further, side-stepping questions of citizen engagement and consent within an epidemiological frame that equates biofortification with initiatives such as water fluoridation: 'The [required nutrients] will get into the food system much like we put fluoride in the water system. It will be invisible, but it will be there to increase [nutrient] intakes'.[3]

This framing implicitly homogenizes 'users' in ways that ignore the diversity of local contexts that shape lived, bodily experiences of health and hunger. At the same time, a standardized approach to target-setting for nutrient levels in biofortified crops, set with reference to a generic understanding of biological 'impact', is conducted in isolation from the environmental and cultural contexts in which such crops might be grown, processed and consumed. Nevertheless, as later chapters reveal, assumptions that such social and ecological diversity can be 'dealt with' further downstream, during the later stages of adaptive and applied research, have been confounded by the diverse and often unpredictable realities of interactions between crops, the environments in which they are planted and the bodies of the people who consume them. At the same time, these findings highlight international biofortification research as an illuminating example of how, as uncertainty increases, the attraction of linear models and reductive analyses that seem to tame uncertainty become all the more powerful (Hacking, 1990).

Focus on Rice: Iconic Crop, Model Cereal

International biofortification research initiatives, including HarvestPlus, cover a range of staple crops.[4] This book has focused on one of these crops: rice. There were a number of reasons for this choice. Rice is widely considered to be the world's most important crop, a view that led the United Nations National Assembly to designate 2004 the 'International Year of Rice' (FAO, 2003). Its significance, however, is complicated by a contradiction: it is both the largest and smallest of cereals – large in terms of aggregate production and consumption, but small scale and highly diverse in practice, with rice markets shaped by 'very strong preferences for different varieties'.[5]

This fragmentation of rice markets has historically deterred private-sector investment, leading rice research and development to concentrate within the public sector. Notably, the International Rice Research Institute (IRRI), founded in 1962 as a 'definitive centre for rice research' and based in Los Baños, was a pioneering venture of the Ford and Rockefeller Foundations and the government of the Philippines, which later served as the prototype for the CGIAR network of crop research centres (Anderson et al, 1991). Even today, IRRI remains the iconic CGIAR 'centre' (Perlas and Vellvé, 1997; Cullather, 2004). This focus on public rice research intensified during the 1980s and 1990s, with the Rockefeller Foundation investing substantial funds, over a period of 17 years, in the development of international capacity in rice biotechnology research (Evenson et al, 1996).[6]

One outcome of the Rockefeller Foundation-funded research was the discovery of the pivotal role of rice as the 'Rosetta Stone' for other (more lucrative) cereals such as wheat and maize, suggesting new scientific and commercial possibilities. The publication of draft rice genome sequences in 2002 by scientists at the Beijing Genomics Institute and the Syngenta Company prompted the following commentary from IRRI's then Director-General:

> If a single plant species were to be voted the most popular by scientists and laymen alike, it would be *Oryza sativa*. Rice, the world's most important cereal crop for human consumption, is the food staple of more than 3 billion people, many of them desperately poor. In addition, rice – like *Arabidopsis* – is a model experimental plant; it has a much smaller genome than those of other cereals and a high degree of collinearity with the genomes of wheat, barley and maize. The blending of the complete *Arabidopsis* and rice genome sequences will forever change the way we approach plant biology research. (Cantrell and Reeves, 2002, p53)

These developments add additional layers of complexity to the politics of international rice research and development in ways that intersect with issues and controversies surrounding the biofortification project. One point of contention has been around the question of whether biofortification research will use conventional plant breeding or transgenic techniques. Early HarvestPlus promotional materials emphasized the use of conventional breeding (HarvestPlus, 2004a, p4) – with transgenic research delayed until the second phase. This phasing enabled promoters to deflect criticisms from the 'anti-GM' lobby and bought time to establish 'proof of concept'[7] for the biofortification strategy before exploring the potential of more controversial technologies.[8] In the case of rice biofortification, however, as Chapter 3 on 'Golden Rice' illustrates, the blurring of boundaries between HarvestPlus and pre-existing transgenic research has made this distinction harder to sustain in practice.

An explicit aim of the Challenge Program experiment was to intensify engagement with the private sector (IFPRI, 2005, p6). However, this was not an aim shared by all CGIAR stakeholders; as demonstrated in 2002 by the non governmental organization (NGO) committee's decision to freeze its membership when the Syngenta Foundation was awarded a place on the board.[9] These developments, together with the controversial appointment of an ex-Monsanto senior executive to lead biofortification efforts (including Golden Rice) at IRRI, raised the spectre of corporate interests in rice biofortification.[10] These controversies highlight some of the tensions within international biofortification research in ways that might not have been so apparent within other HarvestPlus crops, at least within the time frame of this research.

On Researching International Science Policy Processes

The starting point for this book has been to question an accepted model informing the design of programmes such as HarvestPlus – of upstream research as an international public good that can be disseminated through a linear diffusion process. At the same time it has sought to understand the significance of what may be new formations and convergences gathering around the idea of biofortification. While global in concept, the practice of science is always located somewhere, at some time. The approach of this book has therefore been to follow the science policy processes that constitute rice biofortification as a 'global' project, as it is enacted by particular people in particular places. Such an approach calls for a constructionist lens that views knowledge production as located and contingent, and a conceptual and methodological approach that replaces the unambiguous language of problem definitions, authoritative decisions, disciplines and roles that populate HarvestPlus literature with the contingency of frames (Schön and Rein, 1994; Apthorpe and Gasper, 1996), located institutional practices (Schaffer, 1984; Fischer, 1998), epistemic cultures (Knorr-Cetina, 1999), negotiated boundaries (Gieryn, 1999) and actor-networks (Latour, 2003).

This epistemological position locates this research within the methodological tradition of ethnography, originally developed within the discipline of anthropology but now extended to other areas of social science, notably science studies. Strathern notes a particular strength of anthropology as 'taking as *local* those bureaucratic structures, nationalist and internationalist ideologies and claims about universal characteristics that appear everywhere' (Strathern, 1995, p164). In this case, global 'scale' is not taken as given, rather the 'rubric of global and local relations provides the coordinates' for studying specific sets of relations (Strathern, 1995, p164). Tsing has proposed two levels of analysis for researching 'global' processes; first, tracking the rhetorical 'moves' and definitional struggles around claims for global scale, or 'scale making'; and second, tracing the concrete trajectories followed by 'projects', or 'relatively coherent bundles of ideas and practices as realized in particular times and places' (Tsing, 2002, pp472–6).

Here the 'global' research field is conceptualized in a way proposed by Gupta and Ferguson (1997) as a socio-political rather than a geographical space. This notion of 'the field' lies at the heart of a notion of multi-sited ethnography, understood as a conceptual and methodological point of departure (Marcus, 1995), rather than a multiplicity of sites per se. This repositioning is exemplified by new approaches to researching policy processes (Shore and Wright, 1997) that combine attention to processes of knowledge production with an explicit emphasis on the power relations within which knowledge is produced and issues and problems are framed:

> Policies are inherently and unequivocally *anthropological* phenomena in a number of ways: as cultural texts, as classificatory devices with various meanings, as narratives that serve to justify or condemn the present, or as rhetorical devices and discursive formations that function to empower some people and silence others. Not only do policies codify social norms and values, and articulate fundamental organising principles of society, they also contain implicit (and sometimes explicit) models of society. (Shore and Wright, 1997, p7)

The ethnographic study of policy processes 'brings together in one frame of study the social grounds for certain policy discourses and the situated communities affected by such policies' (Marcus, 1995, p100), and therefore represents a point of departure from the traditional role of the social sciences in studying the effects of policies on local communities, within the constraints of a 'resistance and accommodation' framework (Marcus, 1995). In this case, policy processes, rather than their impacts, are the object of analysis and the research field is reconfigured as a 'social and political space articulated through relations of power and systems of governance' (Shore and Wright, 1997, p14). Thus it goes beyond notions of 'studying down' of affected communities, and even 'studying up' of decision-makers at the 'top' (Wright, 1995, p79), to 'studying through'; 'tracing ways in which power creates webs and relations between actors, institutions and discourses across time and space' (Shore and Wright, 1997, p14).

This approach to studying *through* processes in which the 'global' is collapsed into 'local' social and institutional dynamics is key to understanding the ways in which an explicitly centralized programme such as HarvestPlus is always enacted locally, whether in a meeting room in Washington, DC, a laboratory at Cornell or an experimental station in Los Baños. These dynamics are explored in various ways in the following empirical chapters. For example, in Chapter 2 an ethnographic lens is used to highlight processes of knowledge production, network extension and boundary negotiation through which an 'international' initiative around high-iron rice was 'localized', given shape and meaning through a combination of social and political factors that were both quintessentially Filipino, and at the same time illustrative of IRRI's location as 'in but not of the Philippines'.[11]

Tracing actor-networks

A key strategy employed throughout this book, from its original inception, through the conduct of fieldwork to the process of writing can be described as tracing actor-networks (Latour, 2003). Attention to the dynamics of actor-network formation and extension can highlight processes whereby actors become macro-actors and projects and discourses become 'global', rather than taking these as given (Tsing, 2002). Moreover, while focusing on agency and actor perspectives, this is qualified by an understanding of actors' situated agency based on their positioning within networks, which may become fixed for periods of time, but never irreversibly so (Callon, 1991). This in turn depends on the ongoing processes of enrolment of key actors (or non-human agents or 'actants' such as rice varieties or genes, for example) and mobilizations of broader support (Callon, 1986).

This research therefore followed actors, ideas and 'projects' as they moved between CGIAR centres such as the International Food Policy Research Institute (IFPRI) and IRRI, Philippine Rice Research Institute (PhilRice), Cornell,[12] donor agencies and NGOs, universities and research institutions in the Philippines, China and elsewhere. The three empirical 'cases' that constitute the core of this book in fact emerged through the process of field research as 'relatively coherent bundles of ideas and practices' that could be followed as 'projects' as defined by Tsing (2002, pp472–6). The cases do not constitute complete or comprehensive accounts, however, but particular actor-networks created by my own tracing and linking of actors and events through time and space. In particular I followed the practice of 'feeding off uncertainties' (Latour, 2003, p115), as instances where actor-networks become visible, in an attempt to illuminate the 'movement and energy' of network extension (Latour, 2003, p128). The empirical chapters trace actor-networks as they extended in support of the following three 'projects', in each case highlighting the ways in which issues around research organization, interdisciplinary integration and the nature of 'impact' have been framed:

- high-iron research pioneered by scientists at the IRRI and its partners, initially under the CGIAR micronutrients project in the 1990s and later under a regional programme supported by the Asian Development Bank (ADB);
- the 'Golden Rice' project initiated in the early 1990s by scientists at the Swiss Federal Institute of Technology under the Rockefeller Foundation's International Program on Rice Biotechnology (IPRB), subsequently transferred to a public-private partnership with the Syngenta company taking a pivotal role; and
- the first phase of HarvestPlus, one of four pilot Challenge Programs of the CGIAR, launched in 2003 with funds from a range of sources including the BMGF, and its evolving relationship with research activities of the ProVitaMinRice Consortium (PVMRC), under the BMGF's 'Grand Challenges in Global Health' initiative.

By zooming in on critical moments of uncertainty and controversy, these accounts highlight dynamics of network extension, actor enrolment and black boxing. These insights are complemented by attention to the life and career histories of individual actors (Bertaux, 1981), and how institutional and epistemic cultures and framing assumptions have influenced their own particular worldviews and positions over time.

Locations and positioning

The construction of the research field through the theoretical and physical linking of sites raises new ethical challenges, however (Marcus, 1995; Marcus, 1998). It is not a straightforward matter of declaring a position at the start; multi-sited research requires constant reflexive attention to processes of positioning and repositioning during the course of the research. Decisions as to the selection and ordering of sites, the framing of the research on arrival at each site, and the selection of information and insights from one site shared at the next were all part of an ongoing process of 'mobile positioning' (Marcus, 1995). Often it was a matter of responding to opportunities that arose rather than following a carefully made plan. For example, an invitation to the monthly 'spouses' lunch', fortuitously scheduled on my first day in Los Baños, provided an opening for an informal introduction to IRRI life through the eyes of 'IRRI spouses'.

These processes of connection and interpretation were also influenced by how I was received and perceived. My advanced age (for a doctoral student) proved to be an advantage, as did a combination of a familiarity with the natural sciences (the result of previous engineering training) and the status of an interested learner in relation to the biological sciences relevant to this study. Institutional and personal connections, some foreseen, others entirely serendipitous, played a significant role in my ability to gain access to institutions and informants, and so to the course of the research and the knowledge produced. At various times, the reputation of my institutional home, the Institute of Development Studies (IDS) at the University of Sussex and the location of informants on the Sussex campus, enabled me to approach key informants within the HarvestPlus network, and in particular the CGIAR centres, IFPRI and IRRI. However, access to 'daily life' at IRRI – from the research station to staff housing compounds, from meetings and seminars to parties and chats over coffee – so critical to this research, was facilitated largely through a serendipitous series of connections beginning with a fellow doctoral researcher, with whom I had shared similar experiences of development work in Southeast Asia.

While methodologically inductive and open-ended, this is nevertheless positioned research. The strategy of tracing actor-networks across time and space involves a series of choices, and not only about who to interview, how, when and where. It involves conscious (and unconscious) acts of connecting some actors, events, ideas and things, and not others, towards building a coherent, convincing narrative. Coherent, in this case, does not mean comprehensive; what follows is a without doubt a partial account, an outcome of a series of interactions between a researcher and a set of actors. It is also the outcome of a process in which periods of engagement were followed by periods characterized by conscious effort to gain distance, reflect and synthesize. This process helped avoid premature attachment to particular interpretations or orderings, enabling me to continually revisit and refine emerging themes and interpretations and actively seek a range of subject positions and perspectives within a 'constantly mobile, recalibrating practice of positioning' (Marcus, 1995, p113).

This research also raised questions of accountability: to whom should the researcher be accountable in this type of study? It is insufficient to rely on notions of reciprocity and direct accountability to research subjects; however, there are no generally accepted 'norms of partiality'[13] to replace them. The open-endedness of this research agenda, particularly compared to 'scientific research' understood by crop scientists and nutritionists, was interpreted in various ways by different informants, according to how my framing of the research at the time of our meeting resonated with their own view of what constituted the salient issues at that time. The focus on processes rather than impacts, in particular the idea of being critical about processes rather than outputs, was often misunderstood, perhaps because I found it difficult to explain in concrete terms. Before leaving the Philippines I presented my work in progress at an IRRI seminar (Brooks, 2007). Beforehand one of the scientists urged a colleague to attend, since my presentation would remind them 'why we do this work', an expectation I knew I would not fulfil.

One year into the research I responded to a request to write a paper for IFPRI-based members of the HarvestPlus team I had met earlier that year. This created an unexpected opportunity to clarify my position and possibly even influence key actors' thinking during the transition from the first to the second phase of the programme. At the same time I needed to ensure doors remained open to enable me to complete the later stages of my field research. My attention had already been drawn to 'a hegemony among HarvestPlus people ... a tribalism ... If there's dissent you're closed out'.[14] In this case the balancing act that is 'constructive criticism' was not an easy task. In the end, copies of the paper (Brooks, 2006) were shared with a number of HarvestPlus and IRRI staff. In general they would raise specific points that related to 'matters of fact', rather than take issue with my position.

This was also the case at the IRRI seminar, where I expected participants to take issue with a framing of the purpose of my research and the position from which I was conducting it, in particular my proposition that developments in rice biofortification research appear to 'move the locus of decision-making [even] further away from the end user'. However, I stressed that I was not evaluating programmes or technologies or asking 'will it work?' or 'is it any good?' Instead, I was 'interested in how *actors* in various geographic and institutional locations, at different times, *frame* problems and solutions towards addressing [such] questions' (Brooks, 2007, p6, original emphasis). One participant challenged what he saw as the absence of a position – well was I for it or wasn't I? These brief interactions are just an indication of some of the challenges involved in sharing research findings and establishing dialogue across epistemic boundaries. This book charts my attempt to address these questions.[15]

Chapter Preview

This book is organized into seven chapters. Chapter 1 provides an historical and institutional background and context for later chapters. This chapter charts, first, the historical evolution of IRRI as a focal point for international rice research and the CGIAR system within which it is situated. In so doing, the chapter draws attention to elements of continuity and discontinuity, as an institutional framework established in the Green Revolution era has held on to its foundational myths and ways of working, at the same time as transforming itself – or at least appearing to do so – in order to demonstrate its continued relevance in a changing world. Second, this chapter traces a history of successive attempts to bridge the fields of agriculture, nutrition and health in a variety of ways. This discussion reveals that the strategy of biofortification is one of many possible pathways towards achieving such synergies. Its distinctiveness lies in a particular formula within which these elements are conceptualized and causally linked; a formula that is indicative of certain ways of thinking about development, health and the role of science and technology that have converged at a particular time.

Chapter 2 follows the first of three biofortification 'cases', high-iron rice. This initiative attracted funds from the Asian Development Bank (ADB) under a regional nutrition programme at a critical point in time for the promoters of biofortification, ensuring the continuation of research efforts within the CGIAR. With its beginnings in the early struggles to attract recognition and resources, the origins of the iron rice project can be found in holistic 'food systems' (Combs et al, 1996) approaches to 'tailoring the plant to fit the soil' (Bouis, 1995b, p18). Over time, however, this open-ended approach to an evolving, interdisciplinary research effort was narrowed and channelled into a research formula long established within CGIAR breeding centres. Through successive stages, adherence to a hegemonic 'genetics-led' paradigm framed out uncertainties around GxE (genotype by environment) dynamics, streamlined the conceptual and empirical contributions from nutritional research and sidelined questions around the critical role of post-harvest practices in grain nutrient retention.

Nevertheless, the location of the core research network, or 'family', close to their field site and in regular contact with 'actor-subjects', meant that attempts at 'black boxing' (Latour, 1987, p131) these areas of uncertainty were only ever temporarily and partially successful. Upstream and downstream, in this case, were within a couple of hours drive from each other, so that consequences of particular choices were never far away. In time, however, this network dispersed, as its members moved on for further study or relocation to other CGIAR centres. With their replacement by a more 'international' network with its eye upstream, a series of black boxes closed, at least for the time being.

Chapter 3 presents a very different rice biofortification pathway. Following a similar timeline as IRRI's high-iron rice, the Golden Rice story began in a university laboratory in Switzerland. With funds from a Rockefeller Foundation programme which set out to build international rice biotechnology capacity, a research team managed to genetically engineer a precursor of vitamin A into rice, hence its 'golden' colour. The high profile announcement of this discovery, together with a set of claims as to its potential significance, provoked one of the most hotly contested science policy debates in recent years. Less well understood has been the impact of this controversy in shaping the subsequent innovation pathway.

The Golden Rice story provides a vivid illustration of the consequences of the polemicization of a complex problem. The focus on the genetically modified character of Golden Rice has deflected attention away from uncertainties around questions about post-harvest retention and nutritional effectiveness, for example, and towards institutional solutions to anticipated regulatory 'roadblocks'. In this case, the adoption of boundary terms such as 'proof of concept', borrowed from the world of business development, has created a new language with which to obscure and accommodate higher levels of uncertainty.

Yet, despite an emphasis on institutional innovation, on closer inspection the Golden Rice trajectory is essentially a reproduction of a conventional model of linear diffusion (Rogers, 2003), but with an additional dimension – the anticipation of a potentially resistant, rather than a merely passive, public. These dynamics are further complicated in the Philippines where, in the context of IRRI's often unsettled relationship with its host country, international polemics have been reproduced rather than transformed. In this case, the final obstacle of winning acceptance is envisaged as a matter of wearing down irrational anti-GM sentiment – a transformation viewed as isolated from broader questions about the acceptability, within a rice culture, of a variety (or varieties) of such distinctive appearance, texture and taste.

Chapter 4 traces the more recent 'global' convergence around biofortification research. HarvestPlus evolved from the earlier CGIAR micronutrients project and, in the process, it has absorbed a range of regional initiatives within an ambitious global vision, buoyed by the reassertion of the CGIAR's comparative advantage as a generator of international public goods, achievable through a return to upstream research. Repackaged as a Challenge Program, HarvestPlus embodies the tensions inherent in a new way forward identified for the CGIAR, in which a range of interests and agendas are to be reconciled within 'strategic partnerships' and channelled towards 'pro-poor' ends. This shift upstream was coupled with a reconfiguration of interdisciplinarity from earlier food systems approaches towards institution-level partnerships between research organizations representing complementary areas of expertise. However, as this chapter shows, a focus on these upstream partnerships obscures increasingly hierarchical relations further downstream, constraining the development of truly integrated practices.

This chapter focuses on a series of interdisciplinary uncertainties, high-lighting the social dimensions of encounters between different epistemic cultures (Knorr-Cetina, 1999). In the context of evolving internationalized science networks, these encounters reveal asymmetries in the potential of different 'machineries of knowing' (Knorr-Cetina, 1999, p2) to articulate a language of global science. It is these subtle dynamics, rather than pronounce-ments about the importance of thorough nutritional testing or the use (or not) of transgenic techniques, that are shaping the boundaries of biofortification research. Similarly, the boundaries demarcating programmes such as HarvestPlus and Golden Rice are revealed as contingent on the extension of networks through which they are intimately intertwined.

While acknowledging the legacy of earlier initiatives, such as the high-iron rice project, HarvestPlus in practice owes far more to the Golden Rice initiative. In this case a rationale for the roll-out of Golden Rice as a product has been extended to inform impact predictions for a range of biofortified varieties planned under HarvestPlus. While couched in the language of a public health and pro-poor development, HarvestPlus continues this tradition of 'reaching end-users' with predetermined products. This centralized vision relies on a dual construction, combining a notion of 'global' science based on an imagined interdisciplinary consensus and a reframing of impact around constructed 'populations at risk' as aggregates of individualized 'end-users'.

Chapter 5 presents a synthesis of the three cases, revisiting the themes of research organization, interdisciplinarity and understandings of impact. Thereafter, this chapter returns to the initial question that prompted this research: do developments around biofortification tell us something about future directions in the organization and conduct of international agricultural research? What are the implications of these configurations of actors, ideas and resources, now evolving around an implicit formula for 'global science, public goods', in the name of improved human welfare and development? Bioforti-fication, still a young science, provides an illuminating lens through which to question a remarkably resilient set of assumptions linking science and development, and, in particular, the power of 'breakthrough science' to solve what are, ultimately, complex social problems.

Furthermore, in this era of global 'challenges' and a new generation of ambitious, collaborative research arrangements to meet them, what are the likely trade-offs involved in holding together these increasingly complex networks? To what extent can these networks live up to the all-embracing image they project? These questions are particularly relevant as donors find themselves under pressure to disburse larger funds, while streamlining their own operations for greater 'efficiency'.[16] In view of this, the chapter revisits the question of 'upstream–downstream' relations, noting the shifting roles of national partners and increasingly generic notions of 'users', as indicative of where these trade-offs might lie. Finally, the concluding chapter highlights key areas of change which might allow more diverse and context-responsive alternatives to emerge.

1
'Old Lessons and New Paradigms':[1] Locating Biofortification

This book attempts to shed light on an apparent convergence on biofortification as 'a new paradigm linking agriculture, nutrition and health',[2] and in particular on HarvestPlus. It therefore focuses on a point of intersection of two historical paths that have converged and diverged at different times and in different ways in the last 50 years: the evolution of the CGIAR as the leader and exemplar of international crop research; and a succession of attempts to generate synergies between the agricultural sector and nutrition and/or health.

This chapter sets out the historical and institutional context for the biofortification case in terms of where it is located – in the world of international crop research, in particular the CGIAR – and in terms of what it attempts to understand – the meaning and significance of contemporary imperatives to integrate nutrition and health concerns within what has been called a 'new paradigm' for international agricultural research. The institutional focus on the CGIAR reflects its disproportionate influence on 'embodying and setting the standards of professional excellence' in international public agricultural research (Pretty and Chambers, 1994, p195). More specifically, a focus on the world's most important staple crop, rice, has meant that much of the research for this book was conducted in and around IRRI, which, in addition to its role as an international authority on public rice research, retains its symbolic status as the prototype CGIAR centre.

This chapter highlights aspects of the history and culture of the CGIAR, and in particular IRRI, as the embodiment of a set of historically located beliefs and relationships both international in scope and shaped by local dynamics and events, which has tended to generate closure around certain types of pathways and not others. The aim here is to provide an historical foundation for reading subsequent chapters in such a way as to illuminate the particular combination of change and continuity that international biofortification research, in its current form, represents.

International Crop Research and the CGIAR

This section charts the evolution of the CGIAR system as an international network of agricultural research centres. While the system was formally established in 1971, its lifespan is generally considered to have begun with the birth of IRRI, almost ten years earlier. IRRI was, in retrospect, the first CGIAR centre, and has remained the most iconic and controversial (for example, see Perlas and Vellvé, 1997). The template for IRRI and other CGIAR centres, however, was an earlier experiment in agricultural research collaboration. The Mexican Agricultural Program (MAP) was a pioneering programme of US-Mexico government-to-government cooperation, envisioned and led by the Rockefeller Foundation during the 1940s.[3] The programme established the 'Office of Special Studies' (later to evolve into the International Maize and Wheat Improvement Center or CIMMYT, the second CGIAR centre to be established after IRRI), an autonomous institution designed to accelerate progress in agricultural modernization, which, it was anticipated, national institutions would ultimately follow (Perkins, 1997).

From the 1940s onwards, beginning with President Harry S. Truman's 'Point 4' speech and subsequent programme (which later evolved into the United States Agency for International Development – USAID), agriculture came to be seen as an essential element of US foreign policy. This development combined two sets of ideas. The first and most obvious was a product of the Cold War climate and the imperative to contain communism in Asia. Modernization of agricultural production was seen as a way to pacify the countryside and increase food production to meet the demands of increasing rural and urban populations; basically a neo-Malthusian argument (Perkins, 1997, p135). The second was a concern of US business interests, articulated through private foundations such as the Rockefeller and Ford foundations, to create the conditions for the future expansion of trade and investment, which required the integration of developing-country agriculture into capitalist national economies. In 1951 Nelson Rockefeller wrote in *Foreign Affairs* that limitations of the world economy were due to 'under-development', and argued for 'widening the boundaries of US national interest' (Perkins, 1997, p103).

These two strands came together in a formula which Perkins calls 'Population-National Security Theory', which emerged from 'incremental efforts to understand the significance of population growth, destruction of natural resources, world hunger, poverty and political turmoil after the end of the [Second World] war' (Perkins, 1997, p119). The integration of these ideas was at least partly facilitated by the revolving door between the Rockefeller Foundation and the US foreign policy establishment during the formative period of 1945–50 (Anderson et al, 1991, pp22–3). Over time, these ideas solidified into a logic that identified overpopulation as the cause, regional instability threatening US national security as the potential consequence and research in plant breeding as the solution. At this point the foundation's experience with the Mexican Agricultural Program provided the model for a more ambitious programme of intervention in Asia (Perkins, 1997).

IRRI: A 'definitive centre for rice research'

In Southeast Asia, questions about food, population and politics inevitably revolved around rice, as it was by far the most important crop in the region (Chandler, 1992; Anderson et al, 1991) if not the world (FAO, 2003). In the Philippines, for example, both then and now, self-sufficiency in rice equates to national security, and the slogan 'rice is life' is a shorthand term for the many ways in which rice – as plant, food, commodity, symbol – is central to social, cultural, economic and political life (Castillo, 2006; Asia Rice Foundation, 2004). At the height of the Cold War it is not surprising, therefore, that foreign-policy analysts highlighted the strategic significance of enabling countries in the region to generate rice surpluses. In the early 1950s, John King, an agricultural economist from the University of Virginia, wrote a piece on 'rice politics' in which he stated that: 'South and Southeast Asia must be made to realize that increased production and a higher standard of living are possible in their own countries without adoption of totalitarian methods. The struggle of the "East" versus the "West" in Asia is in part a race for production and *rice is the symbol and substance of it*' (quoted in Anderson et al, 1991, p36, emphasis added).

In 1962 IRRI opened its doors as 'the first tax-exempt foundation in the Philippines' (Cullather, 2004, p233), the outcome of a series of negotiations between the Ford and Rockefeller foundations and government of the Philippines. That said, from the outset it was clear that IRRI was 'in but not of the Philippines'. IRRI represented 'a new scientific approach and a new institutional arrangement' (Cullather, 2004, p232), based on a model imported from the US: 'Although by the mid-1960s there were scientists of seven nationalities on the staff, each one had been trained ... in a US university. Most of the first Americans at IRRI recognized that they did not know much about rice or Asia, but they did have a model of an American experimental station' (Anderson et al, 1991, p73).

The vision driving this institutional innovation in its formative stages emerged from a set of ideas about agriculture and development that had been circulating within the Rockefeller Foundation over several years (Anderson et al, 1991). As early as 1951, foundation officials had defined agriculture as 'the application of the principles of biology and other natural sciences to the art of growing food' (Anderson et al, 1991, p32). This definition underpinned a vision of agricultural development assistance in which 'the food problem' was reduced, first to a problem of production, and then to 'yield' as 'an isolable problem' (Anderson et al, 1991, p53) that could be addressed through the genetic manipulation of seeds.

This technology-first approach rested on the twin principles of 'universal' application and 'optimal' conditions: the aim would be to produce plants with universal application, to be achieved by identifying plants that perform best in 'optimal' conditions – which in the case of rice meant flooded irrigation. As Robert Chandler, IRRI's first director-general, remembers:

> The basic problems concerning rice are universal problems, which can be properly attacked in one central laboratory, which would then make the results available to all. Many of the really fundamental physiological, biochemical and genetic problems are essentially independent of geography and are certainly independent of political boundaries; so that these problems could effectively and efficiently be attacked in one central institute. (Chandler, 1992, p2)

With the transplanting of the American experimental station model came the transfer of the 'classic cluster' of agricultural sciences found at research stations in the US – plant breeding, agronomy, pathology, soil chemistry and entomology (Anderson et al, 1991, p73). Other disciplines were added later: agricultural economics (1963), communication (1963) and water management (1968). However, 'all of these new additions needed to prove their value to the classic cluster'. As important was the hierarchy within the classic cluster, 'which placed breeding and genetic manipulation on top' (Anderson et al, 1991, p74), reflecting IRRI's prioritization of 'genetics-led' crop improvement (Anderson et al, 1991, p65).

This account highlights important elements in the early development of IRRI as the earliest manifestation of international collaboration in rice research. While not yet in use, the roots of the application of the term 'international public goods' to agricultural research, now central to the identity of the CGIAR (Science Council, 2006), can be found in these formative elements. The idea of a 'definitive centre' was based on the assumptions about the universality of the knowledge it might produce, assumptions which relied on a highly reductionist analysis of the complexities of food production and access. This idea of universality extended to the apparently unproblematic transfer of an American scientific-institutional model to a foreign country, following the precedent already set by the Mexican Agricultural Program in the 1940s (Perkins, 1997). With this transfer came the hierarchically arranged 'classic cluster' of crop sciences with plant genetics, the discipline best suited to such a reductionist enterprise, taking on the lead role.

The Green Revolution and 'miracle rice'

The Green Revolution 'lies at an important intersection ... between the historiographies of technology and US foreign relations' (Cullather, 2004, p228). The term was first used in 1968 by USAID administrator William S. Gaud and was first debated by the US Congress in 1969:[4] '"Green", of course, was implicitly opposed to "red" and was signalling, like a flag, that social reform was not necessary, since technical means in agriculture (evoked by "green") alone were supposed to solve the problem of hunger' (Spitz, 1987, p56).

There are broadly two meanings of the Green Revolution. The first is a more narrow one referring to 'specific plant improvements notably the development of high yielding varieties (HYVs) of rice and wheat' (Griffin, 1979, p2).[5] In 1970, Norman Borlaug (American botanist, Director of Division for Wheat Cultivation at CIMMYT in Mexico) was awarded the Nobel Peace Prize for 'having set in motion a worldwide agricultural development ... the "Green Revolution" ... Borlaug's "miracle wheat" doubled and tripled yields in a short period of time. Similar increases were soon achieved ... at the International Rice Research Institute (IRRI) in Philippines, with rice' (Glaeser, 1987, p1). The key characteristic of these varieties was their 'semi-dwarf' stature, which allowed the use of increased levels of chemical fertilizers and pesticides. Widespread dissemination of these varieties raised production levels significantly during the late 1960s and into the 1970s, as the following account acknowledges:

> [HYVs] were introduced in several Asian countries in 1965, and, by 1970, these strains were being cultivated over an area of 10 million hectares. Within three years, Pakistan ceased to be dependent on wheat imports from the United States. Sri Lanka, Philippines and a number of Latin American countries achieved record harvests. India, which had just avoided a severe famine in 1967, produced enough grain within five years to support its population. Even after the 1979 drought, grain imports were not necessary. India had become self-sufficient in wheat and rice, tripling its wheat production between 1961 and 1980. (Glaeser, 1987, p1)

A second, broader meaning of Green Revolution refers 'to a broad transformation of agricultural sectors in developing countries, to a reduction in food shortages and undernourishment, and to the elimination of agriculture as a bottleneck to overall development' (Griffin, 1979, p2). It is through this second meaning that the Green Revolution converged with nation building and development objectives of newly independent states towards modernization and industrialization. In South and Southeast Asia in particular, the formula of food self-sufficiency, modernization and technicism found resonance with a 'new generation of populist leaders, whose slogans emphasized developmentalist, rather than redistributionist goals' (Cullather, 2004, p245).

This resonance is exemplified by the nature and timing of the release of IR-8, IRRI's first genetically-improved, semi-dwarf variety, in the Philippines, IRRI's host country. Initial test results showed evidence of yield increases, but the quality and taste were poor, and there were unresolved questions about vulnerability to pests. Despite these reservations, the seed was simultaneously submitted to the seed board for certification and released for preliminary distribution. The seed board resolved its dilemma by awarding the label of 'good seed' while delaying the decision about certification, which was finally granted in 1968 (Anderson et al, 1991, pp64–8; see also Chandler, 1992, p114).

The early release of IR-8 is illustrative of a combination of international and domestic pressures at that time. In India, scientists had produced a steady stream of semi-dwarf wheat varieties during the mid-1960s. This put pressure on IRRI's rice scientists to produce results, pressure that was intensified by donor concerns about an impending food crisis and conference talk among scientists of 'a race against time' (Cullather, 2004, p242). The more immediate pressures, however, were closer to home. In 1962 the Ferdinand Marcos government won the national Philippines election with the slogan 'Progress is a grain of rice'. While the plant breeders urged caution, 'the Philippine press proclaimed the breakthrough had already been achieved' (Cullather, 2004, p243).

The 1965 Philippine election campaign is revealing of the context in which IRRI scientists felt compelled to release their 'miracle rice'. The press announced that: 'spectacular yields were automatic, "lodged into the grain itself – built-in productivity"' (Cullather, 2004, p243). In reality this was far from the case: IR-8 and its successors were part of a *package* that included the chemical inputs necessary to achieve the promised increases in yield (Cullather, 2004; see also Griffin, 1979; Pearse, 1980).

Nevertheless the myth that the yield increases were built into the seed provided a point of convergence between a newly elected, developmentalist regime and IRRI scientists, particularly plant breeders, for whom IR-8 represented 'an unusual capacity to induce peasants, voters and governments to see their situation differently, and to recalculate their interests and allegiances accordingly' (Cullather, 2004, p230). Rafael Salas, charged by Marcos with controlling the distribution of the package of inputs, articulated this recalculation as follows, using the illuminating turn of phrase 'miracle rice *culture*': 'Even if it wasn't such a spectacular producer ... one would advocate pushing miracle rice culture if only to train the Filipino farmer into thinking in terms of techniques, machines, fertilizers, schedules and experiments' (quoted by Cullather, 2004, p244).

This account reveals how stated goals and concerns about scientific excellence, research collaboration and development impact became intertwined and infused with politics at every level. Perkins goes further, suggesting that that the primary purpose of the Green Revolution was not to solve hunger, but to achieve US national security in uncertain times. Meanwhile, Asian governments embraced this particular model of development assistance in pursuit of their own national security concerns (Perkins, 1997). In this respect the Green Revolution was successful; it helped bolster developmentalist regimes committed to suppressing domestic communist movements in Asia, and facilitated conditions for the future expansion of international trade and investment in agriculture (Cullather, 2004). At the same time, it addressed the sovereignty concerns and development priorities of developing countries by enabling them to achieve popularized goals of food self-sufficiency, and generate foreign exchange (Perkins, 1997).

Despite the productivity gains, from the early 1970s, critiques of the Green Revolution began to emerge, drawing attention to the advantage it gave large farmers over small farmers, exacerbating trends towards rural proletarianization and rising inequality (Frankel, 1971; Frankel, 1974; Griffin,

1979; Pearse, 1980).[6] While 'progressive' farmers in favourable environments had prospered, smaller farmers, many of whom struggled to grow sufficient food in diverse, complex and risk-prone environments (Chambers et al, 1989), had not. Pressure mounted for IRRI to move beyond its single-crop focus, towards more holistic, cropping systems approaches and the particular needs of farmers in less favourable environments.

Answering the critics: Expansion or re-orientation?

This section charts concurrent processes through which IRRI, and then later the CGIAR system into which it was absorbed, responded to new levels of criticism and addressed itself to more holistic systems approaches to agriculture. In particular, it highlights how a context of generously funded, rapid expansion made it possible to accommodate apparently contradictory forces in favour of change and continuity, giving rise to a particular way of dealing with change which, this book argues, has repeatedly resurfaced, at critical moments, in the history of the international system.

By 1968 three more international centres – CIMMYT in Mexico, International Center for Tropical Agriculture (CIAT) in Colombia and International Institute of Tropical Agriculture (IITA) in Nigeria – had been established with the support of the Rockefeller and Ford foundations. It was at this point that the foundations saw a need to broaden the funding base for an expanding network of research centres (Chandler, 1992). During 1969–71 a series of four meetings took place at the Bellagio conference centre in Italy (Chandler, 1992, pp159–62), leading to the formation of a Consultative Group on International Agricultural Research (CGIAR), a donor group of 28 members – Food and Agriculture Organization (FAO), United Nations Development Programme (UNDP) and World Bank; Asian, African and Inter-American banks; Organisation for Economic Co-operation and Development (OECD), Canada's International Development Research Centre (IDRC), the Ford, Rockefeller and Kellogg foundations; and 17 industrialized country governments – to support the evolving system of research centres, with its secretariat based at the World Bank.

A Technical Advisory Committee (TAC) 'composed of distinguished agricultural experts from both developed and less developed countries', established to 'advise the CGIAR on priorities in international agricultural research' (Chandler, 1992, p163), would be housed at FAO. While indicative of collaborative effort, Chandler acknowledges the critical role played by Robert McNamara of the World Bank who 'provided the essential impetus to the movement ... envisioned the idea of a CGIAR [and] influenced FAO and UNDP to join as sponsors' (Chandler, 1992, p163). These developments preceded a period of optimism and expansion: 'In the first year of funding, the Group supported five international research centres. By 1976, the network of centres and programs financed through the CGIAR system numbered 11 and financial support had increased fourfold, to $64 million. In 1981, 13 institutes or programs were receiving support from the group, with a combined budget of $145 million' (Chandler, 1992, p165).

In the context of these organizational developments at the international level, IRRI experienced both an extension of funding possibilities and constraints on its earlier autonomy. As one senior scientist reflected in 1978; 'financial decisions were easy to make in the 1960s because we just gathered twenty people in a room and it was done' (quoted by Anderson et al, 1991, p109). Not so in the 1970s with the increasing bureaucratization of life and work at IRRI and the CGIAR system as a whole. At the same time, critiques of the Green Revolution were starting to emerge, putting pressure on CGIAR membership to produce a response (Oasa, 1987). It was in the context of these combined pressures of expansion, bureaucratization and a felt need – for the first time – to *justify* the contribution of international agricultural research that Nyle Brady, on joining IRRI in 1973 as director-general, asked for a review of IRRI's mission.

A new IRRI vision emerged, which combined old and new elements. First, there was consensus around a vision towards 'mission oriented basic research' (Anderson et al, 1991, p79). This element held in tension two potentially conflicting imperatives, which had shaped the original conception of IRRI: the professional aspirations of scientists towards basic research and the mission-oriented approach of the foundations. The second element – the question of how such research should be conducted – proved more contentious (Anderson et al, 1991). This debate crystallized around a choice between two methodological avenues: 'whether to study what occurs on real farms and pursue experiments in those unpredictable conditions (known as the weaker model) or to continue to widen the base of knowledge of rice under more controlled, station conditions (known as the stronger model)' (Anderson et al, 1991, p83).

It was during these debates that a tension at the heart of the 'definitive centre' concept was played out, between the ideas of, on the one hand, universally applicable, optimal technologies achievable though the plot-lab methodological model, and, on the other hand, the diversity and specificity encountered by cropping (or broader farming) systems approaches to on-farm research. These respective positions showed how different approaches carried with them particular beliefs and assumptions of the nature of often taken-for-granted notions of 'good science' and 'impact'. These differences were highlighted by a TAC team reviewing cropping systems research at IRRI in 1976:

> If much of its work was site- and location-specific, asked TAC, how can it be 'generalized to the development of principles'? 'Their role is not merely to analyse traditional practice, but to challenge and change it, the work of an international institute must be capable of extrapolation and impact beyond local test sites. This is the standard against which the cropping systems programs should be assessed over the next five years'. (Quoted in Anderson et al, 1991. p89)

In 1975 IRRI finally ended its rice-naming policy, announcing 'model methodologies' as its new product. A cropping systems division was established,
generating new rivalries between the plant breeders and a new generation of
agronomists (Anderson et al, 1991, pp84–92). These changes reflected debates
at the international level, with the CGIAR as a whole moving towards farming
systems research (FSR)[7] in an attempt to 'understand the cultivator's world'
(Oasa, 1987, p28). However, these changes represented 'an expansion, not a reorientation' of IRRI (Anderson et al, 1991, p84), which failed to disturb the
hegemony of the 'classic cluster' of crop sciences or the position of plant breeding at the top of IRRI's disciplinary hierarchy. In practice, IRRI scientists
maintained the view that 'conditions in farmers' fields were uncontrolled' and
therefore scientific results obtained there were 'of little value' (Anderson et al,
1991, p91). In this respect, they retained the 'implicit belief ... in the existence
of a universal peasant society and a single type of traditional agriculture' (Anderson et al, 1991, p91), a notion fundamentally at odds with the principles of
diversity and site specificity underpinning cropping systems approaches.

Similarly, research into the differential socio-economic impacts of the
dissemination of the high-yielding varieties shifted from an initial question of
the impact on farmer income to a narrower question of impact on yield
(Anderson, 1991, pp94–110). This narrowing of emphasis placed such studies
comfortably within the parameters of the original IRRI vision and at the same
time emptied the research of any political content. This was made possible by
– and at the same time helped to reaffirm – the essential neutrality of
agricultural science. Or, as Oasa succinctly expresses it: 'Research must take a
position of neutrality *precisely because it is not neutral*' (Oasa, 1987, p43,
emphasis added).

In this case the CGIAR expansion and emerging critiques of the Green
Revolution appear as simultaneous and linked processes. IRRI stood at the
juncture of both and carried the main burden of responding to the criticisms.
The IRRI that emerged from its first major review can be seen as the
embodiment of both the 'answer' to the criticisms, in terms of programmes and
methodologies, and of the essential contradictions that these programmes
served to avoid. The struggle that took place between plant breeding and
cropping systems – and the contradictory outcome of a shift in research
emphasis to cropping systems combined with continuity in the hegemony of
plant breeding within the organizational hierarchy – have to be placed within
this context, with IRRI as an arena in which these struggles and contradictions
found mutual accommodation, rather than resolution.

Biggs and Clay (1981, p331) have acknowledged an enduring methodological dichotomy in crop research between 'seeking "widely adaptable"
technologies which can be successfully adopted, even if they are not optimal,
for a range of environments [and] finding or generating the optimal, best
adapted technological solution for specific environments'. They note that, in
practice, 'most formal R and D programmes have either one or the other
orientation, and also that in most cases scientists see the need to follow the

other approach in some of their work' (Biggs and Clay, 1981, p331). In this case, IRRI clearly had the former orientation, while making some attempts to follow the other approach.

The accounts of Anderson and Oasa, however, highlight the organizational politics which, in reality, mediated these negotiations between opposing methodological – and epistemological – positions, located nearer to the 'generic' and 'adapted' ends of the crop research continuum. In this context, without pressure and additional financial support from the wider CGIAR community, IRRI may have been reluctant to invest existing funds in an approach representing such a radical departure from its foundational principles as a definitive centre.

In this case, a combination of pressure and financial support from the wider CGIAR community created a space within which IRRI was able to accommodate cropping systems research without disturbing its internal order and basic philosophy. This pattern of accommodating externally driven programmatic shifts, while maintaining an established institutional structure and set of internal relations, has become deeply embedded in the CGIAR system and its respective centres (Eicher and Rukuni, 2003), as later chapters of this book highlight in various ways.

Synergy or survival? From farming systems research to eco-regional approaches

The 1980s is considered to have been the heyday of farming systems research, within and outside the CGIAR (Zandstra and Taylor, 2006, p5). According to Hubert Zandstra, former head of IRRI's cropping systems division, by the end of the decade 'farming systems research methodology had reached such widespread acceptance that the great majority of international and regional agricultural research centres had adopted FSR as an official part of their programmes' (Zandstra and Taylor, 2006, p7).

Meanwhile, proponents of 'farmer first' (or farmer participatory research) approaches maintained that these developments did not go far enough, since they failed to challenge power-knowledge relations between scientists and farmers so that – though radical in some ways – FSR maintained the conventional assumption that research was something that could only be done by 'professionals'. This assumption was now being confronted with evidence that 'experiences with participatory methods indicate[d] that farmers have a far greater ability than agricultural or other professionals have supposed to conduct their own appraisal, analysis, experimentation, monitoring and evaluation' (Pretty and Chambers, 1994, p195).

The 'farmer first' movement brought together the accumulated experience of initiatives taking place at the margins of mainstream rural development practice over several years, and asserted the value of farmers' local knowledge and its central role in the development of workable, site-specific solutions (Chambers et al, 1989; see also Richards et al, 2009). These developments influenced practice in CGIAR centres such as IRRI, now under the leadership of M. S. Swaminathan, IRRI Director-General in 1982–8 and 'father' of India's

Green Revolution (Seshia and Scoones, 2003). During this time, the 'researcher-designed' cropping and farming systems interventions of the late 1970s and early 1980s gave way to a 'farmer first stage' which incorporated 'farmer-to-farmer training, farmer participatory research, farmer experiments and consideration of farmer practice and technical knowledge' (Fujisaka, 1994, p228).

While these developments suggested the CGIAR might be shifting towards the site-specific methodological continuum of Biggs and Clay, events in the 1990s on the international stage led the CGIAR to choose a different direction. The publication of the Bruntland Report in 1987 and Agenda 21 in 1992 led to an increasing emphasis on integrating sustainability and productivity concerns.[8] 'With these added requirements, the [farming] system became so complex and multi-faceted that new methodologies were required' (Zandstra and Taylor, 2006, p11). During this time the TAC developed the 'eco-regional concept' as a means to focus international research on 'predominant ecologies associated with geographical regions such as the Sahel, the highlands of East Africa or the 'Altiplano' in Latin America' (Zandstra and Taylor, 2006, p11).

The emergence of the eco-regional approach, and its accompanying language of 'big tents', multiple goals and trade-offs, coincided with a funding crisis within the CGIAR system, just as it had absorbed five new centres (Greenland, 1997; Yudelman et al, 1994). The austere financial environment marked a contrast with an earlier time, when the system could simply expand to accommodate and contain an experiment in farming systems research alongside tried and tested methodologies. In contrast, the advent of eco-regional programmes (and a similar development, 'system-wide' initiatives) reflected the twin objectives of programmatic collaboration and cost saving (Fitzhugh and Brader, 2002). It was in this context that, in 1994, the CGIAR's new chairman, Ismail Serageldin, announced a 'renewal of the system' introducing institutional, programmatic and funding changes aimed at transforming the system into a 'fully South-North enterprise'.[9]

According to Zandstra and Taylor, employing this concept 'resolved what had been a major stumbling block to the acceptance of [off-farm FSR] by the TAC – that of site specificity' (2006, p11). However, while the idea of eco-regional approaches encompassed notions of context specificity, in practice the renewed CGIAR was moving 'upstream' towards basic and 'strategic' research (Fujisaka, 1994, p231), which included biotechnology as 'an area that held considerable promise' (Greenland, 1997, p475). This scenario incorporated an untested assumption that the National Agricultural Research System (NARS) would complement this shift upstream by conducting the 'necessary more applied and adaptive research, including on-farm and farmer-participatory research' (Fujisaka, 1994, p231; see also Greenland, 1997).

IRRI, meanwhile, had already undergone its own 'aggressive and painful re-structuring process', beginning in 1988 (Perlas and Vellvé, 1997, p14) with the arrival of IRRI's fifth director-general, Klaus Lampe (1988–95). This was at a time when relations 'at home' were less harmonious than during the Green Revolution era. It was, after all, an enthusiastic *Manila Bulletin* journalist that had first coined the phrase 'miracle rice' (Chandler, 1992, p111) during the election campaign that brought Ferdinand Marcos to power. The Philippines was a very different place in the late 1980s, however, emerging from years of martial law, buoyed by the euphoria of 'people power' that brought Corazon Aquino to the presidency. This period saw a mushrooming of social movements and civil society organizations and activism. For groups whose issue focus was food and agriculture, the time had come to challenge the consensus between elite science and technocracy that IRRI as a 'foreign agency' had come to represent:

> At this time, pressure for change began mounting from both the public and donors. Hostility towards IRRI in its host country grew precipitously in the mid-1980s. An IRRI-instigated survey on the impact of high-yielding varieties on poor farmers led to a major national farmers' conference on rice, where criticism of IRRI was loudly articulated and concerted nationalist efforts to oppose the Green Revolution through farmer-scientist partnerships were launched ... Soon after, public scandal about IRRI's rice blast research being pursued in cooperation with the DuPont Cooperation, among others, nearly bought IRRI to a close down in 1987[10] ... These and numerous other events, involving demonstrations, wide media coverage and even violence, distilled years of farmer and NGO frustration with a foreign agency that had taken a dominant role in directing Philippine agricultural development. (Perlas and Vellvé, 1997, p14)

In light of these more sobering assessments, reports of a 'rejuvenated IRRI' emerging from the restructuring (Perlas and Vellvé, 1997, p14), now focused on a 'Green Evolution: social, political, economic and scientific' (Klaus Lampe, quoted in Perlas and Vellvé, 1997, p33), were met with some scepticism. Meanwhile, following intensive lobbying from agricultural scientists at the University of the Philippines, in 1986 the Philippine Rice Research Institute (PhilRice) was established in Nueva Ecija. As a national institution PhilRice would, it was hoped, direct its attention to the needs and priorities of the Philippines.[11]

These dynamics of hostility and conflict between IRRI and its host country present a very different picture from Lampe's and Serageldin's visions of sustainable 'Green Evolution' and harmonious South-North cooperation. This was in the context of an overarching macro-economic framework, within which structural adjustment reforms were reducing the scope for developing country governments to invest in public institutions, such as those in the agriculture sector (Brenner, 1993; Tabor, 1995). Nevertheless, with agricultural research and development off the radar screen of major donors, what was needed at this time was the projection of a positive image of a renewed and streamlined CGIAR taking on the essential role of strengthening their NARS partners in the South. The problematic nature of this new mission was summarized by one advocacy group as follows:

Another problem sticking the ribs of the CG system is its relations with NARS (National Agricultural Research Systems). The Rockefeller 'pump-priming' vision of international agricultural research centres was one of institutes that would probably phase out in time. They were to be set up to strengthen national programmes by working intimately with them, to the point that IARCs [International Agricultural Research Centres] may no longer be necessary or play a much more subtle role. Thirty years later, however, many NARS are getting weaker and weaker, not stronger. While this is due to factors which are partly outside the control of the CG [CGIAR], it is a major existential problem for the system. In many cases the relationship between NARS and individual centres is one of dependency, love/hate or competition, rather than partnership.[12]

A commentary from the TAC at this time conveys a similar, if more nuanced, picture. In particular the 'basic assumptions' underlying a desired shift in NARS-IARC relations – essential to a new formula combining synergy with site specificity – responded not to an actual state of affairs in respective institutes in terms of capacity and orientation, but to donor and system-level pressures to re-package the CGIAR and its role in a certain way:

> TAC notes with concern the weakened capacity of some NARS engaged in rice research, especially in Sub-Saharan Africa. One of the basic assumptions of the CGIAR is that stronger NARS should take a greater complementary role in the global research system. The solution appears to have more to do with public policy, funding and research management than with the organization of commodity and eco-regional research.[13]

This account points to an emerging practice, within the CGIAR, of articulating certain ideas about scientific excellence and partnership with NARS – constructed within a broader framework of North-South cooperation – in apparent isolation from both the broader international political economic context and realities 'on the ground'.

In a 'meta evaluation' of 'the CGIAR at 31' Eicher and Rukuni (2003) reflect on what appears, in retrospect, to be an unfortunately timed system expansion, forcing cutbacks within core CGIAR centres and falling back on models of research collaboration that could not be realized in practice. Notably, however, the authors distinguish between the financial and political consequences of the CGIAR's foray into natural resource management (NRM), which, they argue, represented an 'egregious management error from a management point of view' (Eicher and Rukuni, 2003, p20). Their reflections on the broader political implications of these decisions, however, in light of contemporary developments within the CGIAR system towards new areas such as public health (as exemplified by the HarvestPlus Challenge Program), are illuminating:

> But when the expansion decision ... is viewed from a political point of view, it is obvious that the system had to join the growing worldwide environmental movement in the early nineties. Besides, the green movement of the early nineties held promise of generating increased financial support in the same way that the Challenge Programs of 2002 are assumed to be attracting new sources of financial support. (Eicher and Rukuni, 2003, p21)

Towards 'strategic research': System-wide initiatives and Challenge Programs

In 1998 Roland Cantrell replaced George Rothschild (1995–8) as director-general of IRRI. A glance over IRRI annual reports before and after this handover is illustrative. Previous annual reports, issued during the mid-1990s with titles such as *Listening to the Farmers* (IRRI, 1996) and *Biodiversity – Maintaining a Balance* (IRRI, 1998), reflected the strands of farming systems and NRM in strategies of the time. Under Rothschild, IRRI had laid off 550 nationally recruited staff, equal to half of its workforce, in the most drastic to date of a succession of staff cuts.[14]

Following Cantrell's arrival, the tone of IRRI annual reports changed markedly, with publication of the 1998–9 report entitled *Rice: Hunger or Hope?* (IRRI, 1999), which seemed to indicate the return of the 'breakthrough mentality' (Oasa, 1987) that had characterized the Green Revolution 30 years earlier. These reports highlighted the potential of iron rice research at IRRI and, further afield, in Golden (high pro-vitamin A) Rice, as well as pioneering, basic research at IRRI into a 'new plant type' (IRRI, 1999; IRRI, 2000b).

Clearly the pressure was on to justify continued funding for international rice research. Cantrell's opening message in the IRRI's 1999–2000 annual report starts with the following questions: 'Why should anyone give money to rice research? What would IRRI achieve with it? What should an "investor" hope to accomplish by donating it?' (IRRI, 2000b, p1) Acknowledging that these are 'perhaps the most important questions facing IRRI and its partners in 2000 – the year of the institute's 40th anniversary and the start of the new millennium', he went on to endorse the decision of the CGIAR 'to rename their donors, *investors*' (IRRI, 2000b, p1, emphasis added). Nevertheless, such rhetoric did not prevent a second round of swingeing cuts at IRRI, following the withdrawal of Japanese funding, in 2002.[15]

During the 1990s the CGIAR had shifted its programmatic and funding priorities from one based on centres to programmes – 'system-wide' and 'ecoregional' programmes or 'SW/EPs' – in an attempt to improve simultaneously collaboration among CGIAR centres, and between CGIAR centres and NARS, and streamline system costs (Fitzhugh and Brader, 2002). By the early

2000s, as these SW/EPs were undergoing assessment, a new funding mechanism called the 'Challenge Program' was being considered as a complement, or in some cases a successor, to the SW/EP.

The Challenge Program took the SW/EP concept a step further by seeking to move beyond the traditional CGIAR mandate and donor base to address 'problems of global importance' that necessitated intensive interdisciplinary collaboration. This shift was described in terms of 'elevating the game' of the CGIAR, and would mean working with new partners 'outside agriculture' and attracting funding from new sources 'which have not traditionally supported agricultural development' (Fitzhugh and Brader, 2002, pp4–6).

These developments took place in the context of yet another major restructuring and reorientation of the system. During the early 2000s the CGIAR underwent a reform process which endorsed a shift upstream, from research *and* development to research *for* development (Science Council, 2006, p6). To oversee the necessary refocusing on a streamlined set of 'system priorities', a Science Council of eminent scientists under the chairmanship of Per-Pinstrup Andersen, a former director-general of IFPRI, took the place of the more broad-based TAC (CGIAR, 2001). The introduction of the challenge fund mechanism and the approval of four pilot Challenge Programs – one of which was HarvestPlus – was one element in this reform process.

With its upstream focus, centralized structure, independent governance, finite time span and opportunities for attracting new partners and funding sources, the Challenge Program mechanism seemed to epitomize, all in a single package, the direction in which the CGIAR needed to move (Science Council and CGIAR Secretariat, 2004). In particular, this model was seen as the type of model likely to engage the key private-sector actors that were now the repository of the resources and proprietary knowledge to which the CGIAR needed access in order to engage in upstream research (IFPRI, 2005, p6). Within the Challenge Program model, CGIAR centres were now recast as brokers or facilitators in heterogeneous international research networks (Rijsberman, 2002). Crucially, these new arrangements provided a framework within which the CGIAR system could sustain its foundational principle of public goods research, while actively pursuing public-private partnership arrangements of various kinds.

Noting significant points of departure in these reforms, towards more 'intense collaborative and interdisciplinary work', Chataway et al (2007, p176) have proposed a rethinking of the concept of 'excellence' in agricultural research as a way forward for the CGIAR. In this case notions of excellence underpinning the role of CGIAR centres as 'centres of excellence' need to be expanded to 'incorporate a broad set of objectives including scientific excellence but also social and economic impacts, the development of collaborative relationships and participative forms, good governance, effectiveness and cost-efficiency' (Chataway et al, 2007, p183). Other commentators go further in opening up the category of scientific excellence itself, using the concept of the 'innovation system' to analyse the distribution of knowledge, expertise and interests within dynamic and evolving innovation systems composed of multiple actors, institutions and stakeholders. In this case, the emphasis shifts from scientific excellence – to be found in 'centres' – to building the innovation capacity throughout the system (for example, see Hall, 2007, pp16–17).

Despite these debates, policy documents emanating from the Science Council (a more influential body that its predecessor, the TAC) suggest that, while endorsing the imperative towards more interdisciplinary and collaborative research, the CGIAR is returning to its roots in basic and 'strategic' research (Science Council, 2006), and with it a return to conventional notions of scientific excellence. The link between basic research upstream and impacts downstream is to be effected through the generation of research outputs as 'international public goods' with potential for wide adaptability and application. As Per Pinstrup-Andersen explained in 2003:

> What is an 'international public good'? Should the Future Harvest centres prioritize the creation of such goods? And, where on the research-development continuum should the CGIAR supported activities be? ... Public goods have two characteristics. First, the use of the good by one individual does not detract from that of another and second, it is impossible to exclude anybody from using the good. A public good is international, if it is of use across country borders. But across how many borders? That is a matter of judgment ... My answer to the second question is YES. Why? For two reasons: First, research that produces private rather than public goods, i.e. goods that can be protected with exclusive property rights, are likely to be produced by the private sector. Second, research results of use to many countries may not generate enough benefits to any one country to warrant national research. *Adding the benefits that several countries can obtain justifies international research.*[16]

Returning to a methodological dichotomy that has been central to recurring debates in agricultural research, between the search for widely adaptable and generic versus site-specific technologies (Biggs and Clay, 1981), Pinstrup-Andersen makes a crucial assertion that a clearer focus on international public goods (IPG) research enables the CGIAR to reconcile these opposing methodological positions:

> National agricultural research systems differ, as do their requests for collaboration with the CGIAR. Some developing countries ask CGIAR Centres to undertake research focused on solving specific national problems because their own research capacity is weak. Others ask for capacity strengthening, while stronger national institutions ask Centres to stop crowding them out of their own countries ... These large differences among countries

call for Centre strategies to be useful to each collaborating country according to its research capacity and interests ... *Is that compatible with a focus on international public goods? Yes, I believe it is* ... In fact, not considering the specific problems and opportunities of each country would limit the impact of CGIAR research. But that does not mean that the CGIAR should do the research and development that is more appropriately done by national institutions. Nor does it mean that the CGIAR should give a reluctant government an excuse not to invest in a national agricultural system. *It does mean that the Centres should plan and conduct research on international public goods that is relevant and useful for application nationally and locally.*[17]

In these statements 'from the Science Council chair', Pinstrup-Andersen appears to have transformed a troubled history into a recipe for complementary contributions and harmonious relations. That this rhetorical move should have provoked so little reaction is revealing of science policy thinking within the CGIAR, itself a reflection of conventional linear models for innovation and policymaking (Clay and Schaffer, 1984a; Godin, 2006).

As with SW/EPs, the responsibility for facilitating local adaptation and adoption rests with NARS and other national and local partners. This is in spite of structural limitations that, as discussed earlier, restrict the capacity of these institutions to take on such a role (Sumberg, 2005).[18] Furthermore, experience to date suggests that the centralized organization and upstream focus may be removing opportunities for farmers and other downstream actors to influence the direction of research. These developments suggest a reversal of efforts over previous decades to balance centrally planned research programmes with attention to site specificity and farmers' knowledge, with a reassertion of the role of the CGIAR centre as the conventional 'centre of excellence'. Reflecting on the current state of farmer participatory research in agricultural research institutes, Ashby notes that in a reformed CGIAR it has been 'molded into a style of technology transfer that uses participatory learning ... to reassert the top down, pipeline model of innovation'. She goes on to assert that:

> Nowhere is this more apparent than in the large scale HarvestPlus and Generation 'Challenge Programs' established by the CGIAR at the end of the 1990s to tackle ambitious plant breeding objectives on a system-wide basis. Driven by what scientists perceive to be their comparative advantage in supplying biotechnology-supported plant breeding solutions to researcher-prioritized problems such as micronutrient deficiencies and drought, these 'mother' programmes define 'baby' farmers as 'customers', and at a strategic level have relegated interaction with farmers to the late stages of delivery of near-finished research products. (Ashby, 2009, p42)

At the same time, however, a central aim of the Challenge Program experiment was to leverage funds from new sources, in much the same way as was envisaged in the 1990s when the CGIAR extended its mandate to include NRM (Eicher and Rukuni, 2003). In initiating Challenge Programs, such as HarvestPlus in particular, the CGIAR has ventured into the domain of human health, an area in which it cannot claim to have expertise (Science Council Secretariat, 2005). In this case, questions raised by Chataway et al (2007) and Hall (2007) become all the more pertinent and highlight inherent tensions between CGIAR centres' habitual role as definitive centres and their new role as 'brokers' in networks in which they must look to others as the experts (Rijsberman, 2002).

How these tensions are resolved (or not) is explored in various ways through the empirical cases in this book. Reflecting on the lessons drawn from their evaluation of 'the CGIAR at 31', Eicher and Rukini (2003) draw attention to the pitfalls of ambitious 'global' agricultural research agendas and to the capacity of individual CGIAR centres to appropriate such agendas, and the resources they bring, in such a way as to protect rather than challenge established practices: 'The introduction of Global Challenge Programs has raised questions about the wisdom of setting global research agendas for agriculture... There is also a concern that Centre Directors can block change by trying to use the CPs [Challenge Programs] to raise new resources while holding onto old mandates and activities' (Eicher and Rukuni, 2003, p24).

Earlier sections of this chapter have highlighted this capacity of the CGIAR, and its component 'centres', to appear to accommodate change while 'holding on to old mandates and activities'. In successive phases of development, these institutions have adopted the language and, to some degree, the practices of holistic farming systems research, ecological sustainability and North-South partnership. With biofortification research, the CGIAR had identified a 'new paradigm' within which to demonstrate its continued relevance.

This was not the first time that the CGIAR and agricultural research community had attempted to establish connections with issues around nutrition and health, however. The next section attempts to place these contemporary developments in the context of previous efforts by the CGIAR and others to bridge agriculture, nutrition and health: from the days of the Green Revolution to HarvestPlus.

Pathways Linking Agriculture, Nutrition and Health

In January 2008, more than 40 years after the Green Revolution began, an international team of public health scientists coordinated the publication, in the influential *Lancet* medical journal, of a series of articles documenting the continuing high prevalence of maternal and child malnutrition in many parts of the world. This situation persists despite the availability of a range of proven, effective interventions and a comprehensive array of international institutions[19] involved in nutrition-related policy and programmes (Horton, 2008, p179).

In contrast to agriculture, the field of international nutrition has been subject to constant changes in definition, identity and institutional base, and this has been reflected in a plurality of approaches to nutrition policymaking and programming (Gillespie et al, 2004). As highlighted in one the *Lancet* series papers:

> Over the past fifty years, countries of low and middle income have witnessed many changes in international thinking with regard to strategies for reducing malnutrition, driven by a variety of forces beyond their control. During the past half century we have had the protein era, the energy gap, the food crisis, applied nutrition programmes, multi-sectoral nutrition planning, nutrition surveillance, food insecurity and livelihood strategies, and the micronutrient era, among others. These fashions do not generally end abruptly, instead bleeding into one another and leaving relics in place within countries and organizations long after their heyday has passed. Only rarely do these fashions reflect changes in the nature of nutrition problems on the ground in poor countries. (Bryce et al, 2008, p1)

The message here is that, while international nutrition has engaged with a plurality of paradigms in the course of the last half century, this has not necessarily been a reflection of greater context-responsiveness, but rather an outcome of inconclusive debate and fragmentation at the international level, so that comprehensiveness has been achieved at the expense of clear prioritization and focus. In practice, these competing paradigms have coexisted, supported by their respective champions, diverting attention from local successes and opportunities for 'scaling up' workable solutions (Bryce et al, 2008). It is in this context that universal solutions most able to bypass institutional shortcomings through top-down vertical programmes, such as salt iodization and vitamin A supplementation, have emerged as the international nutrition success stories of our day (Bryce et al, 2008).

Nevertheless, the role of malnutrition as a 'risk factor' in relation to morbidity and mortality, particularly for women and children in developing countries, is well known and was highlighted in the Global Burden of Disease study commissioned by the World Bank in the early 1990s, when it turned its attention to 'global health' (Murray and Lopez, 1996). Similarly, a persuasive case can be made for investments in nutrition in developing countries as a means to achieve a set of measurable targets, incorporated into the overarching global framework of the Millennium Development Goals, within which development interventions are now routinely explained and justified: '80% of the world's undernourished children live in just 20 countries. Intensified nutrition action in these countries can lead to the achievement of the first millennium goal ... [halving poverty and hunger] and greatly increase the chances of achieving goals for child and maternal mortality (MDGs 4 and 5)' (Bryce et al, 2008, p1).

Agriculture-nutrition-health linkages: A brief history

It was in light of a disappointing mid-term review of progress towards the MDGs[20] in 2005 that international agencies decided to look again at possibilities for improving nutrition and public health outcomes through improved linkages between agriculture, nutrition and health. Major studies were undertaken by IFPRI (for the CGIAR) and the Agriculture and Rural Department of the World Bank (Hawkes and Ruel, 2006b; World Bank, 2007), synthesizing analytical and empirical contributions from the previous two decades (for example Pinstrup-Andersen, 1981; Lipton and de Kadt, 1988; Kennedy and Bouis, 1993; Haddad, 2000; Johnson-Welch et al, 2005).

Efforts to link agriculture, nutrition and health can be grouped into the following phases. The 1960s was dominated by the thinking surrounding the Green Revolution, which tended to exaggerate the existence of agriculture-nutrition linkages (von Braun, in Kennedy and Bouis, 1993, pv), by assuming that nutritional needs would be met by plugging the 'food gap' with higher production levels. This view continued into 1970s debates about 'food security': 'Programmes operated under an assumption that agriculture's primary role in this equation was to address protein-energy deficiencies by increasing food production and reducing prices' (World Bank, 2007, p8).

Towards the end of the decade, however, publications on agriculture-nutrition linkages became critical of assumptions that production increases at the national level would automatically translate into improved nutrition (Kennedy and Bouis, 1993; World Bank, 2007b). In a report commissioned by the World Bank, Per-Pinstrup Andersen (1981) argued that if agricultural development was to contribute to improved nutrition, 'nutritional aims would have to *explicitly* incorporate into agricultural production decisions' (quoted by World Bank, 2007, p8, original emphasis).

During the same period, an example of a 'relic' from a former nutrition era[21] was the Quality Protein Maize (QPM) breeding programme at CIMMYT. Work began in the early 1970s to develop high-lysine maize varieties, in the CGIAR's earliest experience with what would later be called 'biofortification'. The breeding programme met a major difficulty, however; varieties with high lysine were low-yielding. Ultimately, the programme was largely abandoned when the emphasis within nutrition shifted away from protein, leaving the plant breeders involved somewhat disillusioned. When the idea of breeding for nutritional traits arose again in the 1990s – this time for micronutrients – the QPM legacy was to prove a major obstacle to its acceptance by CGIAR scientists who considered it to have been 'a major misallocation of resources' (Bouis, 1995a, p5; Bouis et al, 1999).

By the 1980s, influenced by Amartya Sen's work on famines and entitlements (Sen, 1981), the emphasis had shifted to 'how agriculture could help improve nutrition by increasing income among agricultural communities' (World Bank, 2007, p9). In the 1980s there was increased recognition of the multiple pathways linking agriculture and nutrition, in particular:

(1) Increased incomes and lower food prices, which permit increased food consumption; (2) effects on the health and sanitation environment at the household and community levels, which may increase or reduce morbidity; and (3) effects on time allocation patterns, which may increase or decrease time spent on nurturing activities – time that is often related to women's control over household income and is an important determinant of women's nutritional status. (Kennedy and Bouis, 1993, p2)

Studies published in the 1990s following shifts to agricultural commercialization highlighted an important finding: that increases in income 'did not substantially improve child nutrition status, leading to the conclusion that *income alone could not solve malnutrition*' (World Bank, 2007, p9, emphasis added). In the 1990s, a series of international conferences[22] generated a degree of consensus – in principle if not always in practice – around UNICEF's 'food, health and care' framework which incorporated these non-food factors affecting nutritional status, in terms of access to healthcare facilities and conditions of environmental health, and the gendered dimensions of intra-household resource and time allocation and their effects on maternal health and childcare practice. Nevertheless, an enduring food-first bias continued guide to nutrition policymaking in many countries (Pelletier, 1995, p295).

An important contribution to understanding broader agriculture-health linkages was made in the late 1980s in a study commissioned by the WHO (Lipton and de Kadt, 1988). This report addressed the questions about 'the totality of the effects of agriculture on health' (Lipton and de Kadt, 1988, p7) in an attempt to bring health ministries into decision-making about agricultural projects and policies. This study went beyond analysing agriculture-nutrition linkages and investigated the impacts on health of agriculture as a complex and multi-factorial process. As the authors explained: 'The agriculture-nutrition chain constitutes the main set of links between farming and health. We know how man-made decisions – and indecision – about farming affect health. Yet health considerations play little or no part, in most countries, in decisions either by farmers about production, or by government about agricultural projects and policies' (Lipton and de Kadt, 1988, p7).

However, this attempt to expand the role of national health ministries in agriculture was eclipsed, in the wake of widespread structural adjustment reforms (in which the scaling back of national health systems mirrored the experience, discussed in the previous section, within NARS), by a 'paradigm shift' from international health to global health, which has been defined as follows: 'The US Institute of Medicine in 1997 defined global health as "health problems, issues and concerns that transcend national boundaries, may be influenced by circumstances or experiences in other countries, and are best addressed by cooperative actions and solutions"' (Department of Health, 2007, p16).

This shift was marked by the publication in 1993 of the World Development Report, *Investing in Health* (World Bank, 1993), and the accompanying analytical framework provided by the Global Burden of Disease Study (GBDS) (Murray and Lopez, 1996). This study set out to provide 'a standardized approach to epidemiological assessment' through the employment of the disability-adjusted life year (DALY) as a 'a standard unit' for assessing the relative burden of diseases and associated 'risk factors' (Murray and Lopez, 1997, p1436). From this analysis, malnutrition emerged as 'the risk factor responsible for the greatest loss of DALYs ... worldwide' (Murray and Lopez, 1997, p1440).

This framework has been criticized for its overemphasis on epidemiological analysis and an ahistorical perspective which neglects an 'existing legacy of health care provision, infrastructure and power relations' (Barker and Green, 1996, p181). Furthermore, the DALY, despite its neutral framing, speaks to contemporary neoliberal sensibilities by 'reflecting the desire to formulate a single index to express the burden of disease on *lost economic potential* ... rather than the burden of disease on the individual as a whole' (Cohen, 2000, p523, emphasis added). This tension between 'cost-effective versus the equitable application of health interventions' led the WHO to express reservations about the advisability of using the DALY for health planning and resource allocation (Gwatkin, 1997, p141). Nevertheless, this framework and its institutional sponsor, the World Bank, has been highly influential in shaping evolving ideas and priorities around 'global health' throughout the 1990s and beyond.[23]

International nutrition and the shift to micronutrients

During the 1990s the international nutrition community focused increasingly on the problem of micronutrient deficiencies. This was not wholly new: UNICEF had promoted salt iodization since the 1960s and in the 1970s research findings had demonstrated effects of iron deficiency anaemia on 'productivity and cognitive function' (Gillespie et al, 2004, p96). Since the mid-1980s, however, evidence had gathered around the measurable effects of micronutrient deficiencies (Gillespie and Mason, 1991; Bhaskaram, 2002) and their economic consequences (Horton and Ross, 2003). In particular, high-profile findings on vitamin A and immunity indicted that its provision 'had benefits beyond prevention of known deficiency diseases; it also affected mortality from other causes. This research altered the cost–benefit calculation for vitamin A intervention' (Gillespie et al, 2004, p82).

This new emphasis on micronutrients lent itself to an increasingly dominant, 'goal-oriented' approach to nutrition, monitored by international coordinating committees established for vitamin A, iron and iodine (Gillespie et al, 2004), giving rise to a series of vertical programmes. These high-profile campaigns nevertheless coexisted with the continued espousal, by UN agencies and NGOs, of UNICEF's 'food, health and care' framework. Over time, vitamin A supplementation and salt iodization have come to be regarded as islands of success in an otherwise gloomy picture of international nutrition (Bryce et al, 2008). In contrast, the persistence of the most serious – and complex – micronutrient problem worldwide,[24] that of iron deficiency anaemia, exemplifies the contin-ued 'failure of integration'[25] that vertical approaches attempt, not always successfully, to bypass.

The goal-oriented approach is consistent with a 'fixed genetic potential view' of malnutrition, in contrast to the more context-dependent 'individual adaptability view' (Pacey and Payne, 1981, pp37, 49). While the former lends itself to universalized remedies, the latter recognizes malnutrition as 'specific to particular localities, with characteristic agricultural ecologies and work patterns'. Unlike the former view, which treats standardized measurements, such as those based on height, weight and arm circumference, as direct indicators of malnutrition, in the individual adaptability view such measures can only provide an indirect and partial proxy for the nutritional status of particular people in particular places. Clearly, the genetic potential view is more amenable to a 'global', goal-oriented approach to nutrition, allowing the standards of body size and food intake of 'well fed' and 'healthy' populations in the North to serve as the standard to be achieved elsewhere:

> The basic premise here is that there is an optimal or preferred state of health, fixed for each individual, and determined by his or her genetic potential for growth, resistance to disease, longevity, and so on. It is assumed that everyone could and should achieve their full potential and that malnutrition starts as soon as there is any departure from the preferred state. We cannot measure human genetic potential, however, so it is further assumed that the standards of body size and food intake of 'well fed' and 'healthy' populations approximate to this optimum. (Pacey and Payne, 1981, pp38–9)

These two approaches to interpreting malnutrition, and their respective positions on its measurement and treatment, mirror a similar dichotomy within agricultural research and technology development between orientations towards universal applicability or site specificity (Biggs and Clay, 1981). Within international nutrition, however, as noted by Bryce and his colleagues (2008), different nutrition paradigms overlap, with new approaches pursued alongside 'relics' left over from a previous era. In the 2000s, for example, the FAO was successful in securing a consensus around 'The Right to Adequate Food'[26] as an advocacy platform for a 'rights-based' approach to nutrition. This agenda exists alongside UNICEF's 'food, health and care' framework. In practice, however, international attention and resource allocation has followed the 'goal-oriented' model, based on a 'fixed genetic potential view' of malnutrition. As a consequence, international nutrition programming favours supplementation and industrial food fortification approaches offering economies of scale,[27] over small-scale, community-based initiatives oriented towards achieving lasting change in diets and practices (Delisle, 2003).

Micronutrients and the CGIAR

The shift in the international nutrition agenda towards micronutrients and global goals has been reflected in new approaches to linking agriculture, nutrition and health. In 1993, IFPRI received funding from the USAID Office of Nutrition to identify ways in which the CGIAR 'might undertake to join other international and national organizations in the fight against micronutrient malnutrition' (Bouis, 1995a, p11). Howarth Bouis, an economist at IFPRI with a background in food price analysis, took up the challenge and travelled to the CGIAR breeding centres in different parts of the world in an attempt to persuade plant breeders to investigate the feasibility of breeding for nutritional traits.

However, plant breeders, weighed down with a plethora of breeding objectives, and mindful of the QPM experience, were unenthusiastic.[28] The breakthrough came with the discovery that research of a similar nature was ongoing at the Cornell-based Plant, Soil and Nutrition Laboratory (PSNL)[29] and the Waite Agricultural Research Institute of the University of Adelaide. In 1994 a workshop bringing together these researchers, along with CGIAR plant breeders and selected nutritionists, generated a series of papers as a starting point for a study into the feasibility of developing micronutrient-dense staple crops (Behrman, 1995; Calloway, 1995; Graham and Welch, 1996).

On the strength of these findings, a proposal was submitted to the CGIAR Technical Advisory Committee (TAC) as a system-wide programme. This proposal was rejected, however, with 'questions raised concerning the comparative advantage of the approach' (Zimmermann, 1996; Van Roozendaal, 1996, pp73–4). Nevertheless, Bouis was determined to proceed and the project, now entitled the 'CGIAR micronutrients project', continued, albeit with limited funds (Bouis, 1996a).[30] Project results were presented, alongside similar and complementary research, at a seminar hosted by IRRI in 1999.

Entitled 'Improving Human Nutrition through Agriculture',[31] this seminar represented a landmark attempt to facilitate interdisciplinary dialogue between plant breeders and nutritionists. In its approach, biofortification represented one of two 'camps' regarding the conceptualization of agriculture, nutrition and health – those attempting to a build holistic framework as a starting point (Lipton and de Kadt, 1988; Haddad, 2000; Hawkes and Ruel, 2006b) and the biofortification approach as a specific intervention and a potential platform for broader collaboration.[32] The following discussion explores why this second approach caught the imagination of decision-makers within a reformed CGIAR and new donors mobilized through the Challenge Fund mechanism.

In the 2000s, the combination of an intensified focus on the MDGs, another wave of reforms within the CGIAR and the arrival on the scene of an influential new private philanthropist created the conditions for a new set of possibilities to emerge. The benefits of investments in nutrition, in particular micronutrient interventions, are increasingly framed in terms of relative cost-effectiveness on a global scale, with the MDGs as the definitive framework for understanding and achieving development 'impact' (Darnton-Hill et al, 2002;

Horton and Ross, 2003; SCN, 2004a). In this case, the conceptualization of the causal relationship between malnutrition and poverty has undergone a reversal since the multi-sectoral national campaigns of the 1970s and 1980s, in which nutritional improvements were linked to broader socio-economic conditions (Tontisirin et al, 1995). Today, nutritional improvements are considered necessary as inputs that will enable the poor to work their way out of poverty. As Jeffrey Sachs, the director of the UN Millennium Project, argues:

> Besides being a goal itself, nutrition is critical to achieving other MDGs. Undernutrition contributes to dysfunctional societies with individuals too weak, too vulnerable to disease, and too lacking in physical energy to carry out the extraordinarily laborious tasks of escaping the poverty trap. Malnutrition and hunger feed directly into ill health and poverty. Lack of nutrition means children cannot concentrate adequately in schools, compromising efforts to achieve universal education. (SCN, 2004b, p7)

These shifts have profound implications for the way in which 'development' and, by extension, notions of 'impact' are imagined. As Saith (2006) has argued, the MDG framework is not as 'global' as it appears and the treatment of malnutrition within this framework illustrates this point. In particular, a 'fixed genetic potential view' is taken, accepting nutritional patterns in the North as the global standard, while locating the problem of deficiency (as measured against this standard) in the South in the bodies of those individuals who constitute 'populations at risk'.[33]

This repackaging of nutrition as a particular kind of 'global' problem was given a further boost in 2004 when the Danish Environmental Assessment Institute hosted a high-profile conference entitled the 'Copenhagen Consensus'. Based on a series of 'Challenge Papers' on ten global development problems, including one on 'malnutrition and hunger' (Behrman et al, 2004), an 'expert panel' of eight economists ranked a list of projects or 'opportunities' according to their cost-effectiveness. In this exercise, 'providing micronutrients' was ranked second only to 'control of HIV/AIDS'. As an illuminating comparison, the Kyoto Protocol was categorized as a 'bad project' and ranked 16th.[34]

The rise of 'goal-oriented nutrition' (Gillespie et al, 2004) therefore represents a convergence around certain ways of thinking about nutrition, health and development, which reduces development to the economic realm and differential experiences of health and well-being to those factors that register in standardized epidemiological analysis. This type of reasoning resonates with a worldview held by key decision-makers within a new generation of post-corporate, private philanthropic foundations – led by the Bill and Melinda Gates Foundation (BMGF) – whose 'Silicon Valley' model for modern philanthropy is increasingly heralded as the way forward.[35] The turning point for biofortification research within the CGIAR came when, within the same year, a proposal was approved by the Science Council for a

Biofortification Challenge Program (subsequently renamed HarvestPlus) and, on the strength of this, the BMGF agreed to become its largest donor.

In an IFPRI policy brief on evolving agriculture-nutrition linkages in a changing world, Hawkes and Ruel highlight the continued relevance of old lessons as well as new paradigms (Hawkes and Ruel, 2006a). Since the 1960s, these linkages have been conceptualized, in different eras, via aggregate food production (1960s–1970s), household income (1980s–1990s), intra-household time allocation and women's empowerment (1990s–2000s) and, more recently with biofortification, through adjustments to the nutrient content of food crops. Furthermore, initiatives launched in these respective eras have to be placed in a context of attempts to go beyond the more obvious agriculture-nutrition connection and understand the broader implications of all aspects of the agricultural process for human health (Hawkes and Ruel, 2006b; Lipton and de Kadt, 1988). In light of this long and varied history, the strategy of biofortification represents just one of a series of attempts to capitalize on the 'agriculture-nutrition advantage' (Johnson-Welch et al, 2005).

'Old Lessons and New Paradigms'

The significance of HarvestPlus is not that it attempts to link agriculture, nutrition and health, but that it does so within a formula that appealed to a newly instated Science Council steering a new wave of reforms through the CGIAR, and to decision-makers within the acknowledged leader of a new generation of private philanthropists, the Bill and Melinda Gates Foundation. And, while the international nutrition community had yet to endorse the biofortification approach, in principle it did not disturb the paradigm that currently dominates international nutrition programming. In this context, biofortification combines, within one memorable package, the ideal of 'break-through' science as a global public good, the potential of new institutional configurations to attract resources and the promise of remedying one of the world's most emotive and preventable ills, malnutrition.

In contrast to earlier approaches to integrating agriculture, nutrition and health issues, activities and sectors, however, the biofortification approach requires the development of an integrated science. Notably, breeding varieties for nutritional traits involves nutritionists, rather than plant scientists, setting objectives for plant breeding. These evolving interdisciplinary relations raise new challenges for the organization and practice of science. This book questions this 'unique' aspect of biofortification research from two directions. First, it raises questions about the ways in which research collaboration between various types of scientific institutions is organized. As a Challenge Program, HarvestPlus was both 'networked' and centralized, combining elements of 'partnership' and hierarchy, though the former was emphasized over the latter. One entry point for this book, therefore, has been to open up the notion of research partnership and explore, through a series of empirical cases, how these partnerships have played out in practice.

As Ashby (2009) has noted, the centralized, technology-driven design of Challenge Programs effectively reduced the space for farmers to participate in shaping the technologies for which they were now cast as 'consumers' at the end of the technology pipeline.[36] This book explores different biofortification pathways before they had reached farmers and, in so doing, analyses relationships between CGIAR centres and national institutions, including NARS. In particular, did contemporary developments continue patterns observed during the 1990s, when a language of partnership obscured unequal and sometimes conflict-ridden relations (Perlas and Vellvé, 1997),[37] or was the era of the Challenge Program a point of departure in international research collaboration?

Second, this book explores interactions between scientists of different disciplines. Biofortification created new interdisciplinary connections – between plant breeders and nutritionists, for example – as well as challenging the arguably more settled relationships among the crop sciences in new ways. Starting from an assumption that biofortification relies on the successful integration of a range of disciplinary perspectives, both in 'strategic' research upstream and adaptive and applied research downstream, this book has zoomed in on a series of interdisciplinary encounters at key moments in the journey of biofortification research, from its modest beginnings in the early 1990s onwards. It explores the extent to which genuine interdisciplinary integration was taking place. In particular, to what extent did the new requirements of biofortification research challenge – and succeed in shifting – the disciplinary arrangements exemplified at IRRI by the resilience of the 'classic cluster' of crop sciences, led by plant breeding, in the face of pressures to change (Anderson et al, 1991)? Or did the availability of new funds made available through Challenge Programs such as HarvestPlus present CGIAR centres with a window of opportunity for 'holding on to old mandates and activities' (Eicher and Rukuni, 2003, p24)?

Third, this book takes the question of impact – how it is understood and framed – as another key entry point. Both the CGIAR Science Council and the BMGF emphasize the importance of achieving and demonstrating impact in the context of the MDG framework. Similarly, both favour investments in basic and strategic research upstream, assuming that such investments will generate research outputs as international public goods that will prove widely applicable. While the biofortification research pathways followed by this book had yet, in most cases, to reach farmers and consumers, this book has explored the way in which various actors involved in biofortification research have dealt with the multitude of uncertainties about the efficacy, effectiveness and – ultimately – the acceptability to farmers and consumers of the outputs of biofortification research.

HarvestPlus, with its orientation towards public health in general and micronutrient malnutrition in particular, ticks all the right boxes on the MDG checklist. At the same time, however, the field of international nutrition remains a highly contested one, always subject to shifts in new directions. Furthermore, publications such as the *Lancet* series on under nutrition (2008) and the World

Bank book *Repositioning Nutrition as Central to Development* (World Bank, 2006) advocated the prioritization of the nutritional 'window of opportunity' represented by pregnant women and infants under two years. It is well known that the special requirements of these groups call for concentrations of micronutrients that are unlikely to be offered by biofortified staples, a limitation that was acknowledged in the first external review of the HarvestPlus programme.[38] In light of these often contradictory influences, the chapters that follow revisit these questions and attempt to throw light on the potential significance of international biofortification research and HarvestPlus.

2
Building the Argument:
The Case of Iron Rice

Introduction

This chapter traces the history of an early experiment in plant breeding for enhanced micronutrient density. In the mid-1990s plant breeders in IRRI's salinity tolerance breeding programme turned their attention to the possibility that varieties bred for tolerance in unfavourable environments may contain higher levels of trace elements such as iron and zinc in the grain. This coincided with increased political attention to the problem of micronutrient malnutrition in the Philippines, particularly iron deficiency, following the international 'Ending Hidden Hunger' conference;[1] and concerted efforts by Howarth Bouis of IFPRI to generate interest within CGIAR centres in plant breeding for nutritional traits.

These and other connections led to the establishment, in 2001, of a regional iron rice biofortification programme, funded by the Asian Development ment Bank (ADB), based at IRRI and the Institute of Human Nutrition and Food (IHNF) at the University of Philippines Los Baños (UPLB) and involving national research institutions in four countries.[2] This account focuses on events as they played out in the Philippines, in the relations and interactions within a research 'family' linking IRRI, IHNF, PhilRice (the NARS in the Philippines) and ten Catholic congregations in and around Metro Manila. Also included in the family were a core group of researchers, which included Bouis and scientists with whom he had established strong connections in the early years of the CGIAR micronutrients project based in the United States (Cornell) and Adelaide, Australia.

The research produced findings endorsing the 'proof of concept'[3] for iron biofortified rice, which played a critical role in making the case for a larger, international biofortification programme. In addition, for the first time in the Philippines, a rice variety was approved and released on the strength of its nutritional characteristics. However, just as these accomplishments were acknowledged, they were already beginning to unravel and, in the process, a multiplicity of interpretations emerged of the results and their significance, and what should happen next.

This is an account of three concurrent and interwoven stories. The first is one of relatively open-ended processes of interdisciplinary scientific enquiry by a close-knit network of researchers that eschewed myths of linearity in scientific research, acknowledging multiple uncertainties and multi-dimensional research relations and serendipitous discoveries. Second, this is at the same time a story about the development of a product: as research progressed, so too did the process of submitting the experimental material used in the research for assessment as a rice variety for national release. In the process, discussion of 'proof of concept' study findings became conflated with an assessment of varietal performance and viability. Over time, these debates grew increasingly inward-looking, failing to take account of a broader national context in which rice politics had taken an altogether different course.

Third, this is a story of legitimation. For members of the international networks the iron rice research delivered the sought after 'evidence' in support of their argument for an expansion of biofortification research and programming. In the process of constructing this argument, however, critical questions and uncertainties were black-boxed or framed out of debates at critical moments, limiting opportunities to learn from this experiment in interdisciplinary enquiry, with implications for the design of later biofortification initiatives.

A Win–Win Proposition: Nutrition and Yield

In the early 1990s Howarth 'Howdy' Bouis, an economist in the Food Consumption and Nutrition Division (FCND) at IFPRI, began promoting the idea of breeding crops for higher micronutrient density as a cost-effective strategy to reduce global levels of 'hidden hunger' (Bouis, 1995a, p11; Bouis et al, 1999; Hilchey, 1995). As discussed in Chapter 1, with initial start-up money from USAID, Bouis visited CGIAR breeding centres in search of plant breeders willing to commit to such an effort. Plant breeders, however, were unenthusiastic, considering themselves already overburdened with a multiplicity of breeding objectives.[4]

This initial response was also influenced by an early experience with plant breeding for nutritional traits within the CGIAR system: Quality Protein Maize (QPM), at the tail-end of the 'protein paradigm' in international nutrition. Work began at CIMMYT in the early 1970s to produce high-lysine maize varieties. The breeding programme met major difficulties, however; varieties with high lysine were low-yielding (although later efforts have produced higher yielding, high-lysine varieties (Cohen, 1995). Eventually the programme was sidelined (though not completely discontinued), as the emphasis within international nutrition shifted from protein to calories, then later to micronutrients.

When the idea of breeding for micronutrients was first floated, the QPM experience was to prove a major obstacle to its acceptance by CGIAR scientists

(Bouis, 1995a, p12). They assumed that, as with the QPM experience, 'there would again be a trade-off between plant yield and nutritional value or that, at best, there would be no correlation with yield and that adding an additional breeding objective (nutritional quality) would slow down the overriding breeding objectives of higher and more stable crop yields' (Bouis, 1995a, p12).

A food systems approach

A breakthrough came when Bouis met Ross Welch, a scientist at the Federal Plant Nutrition and Soil Laboratory (PNSL), based at Cornell University:[5] 'The PSNL, established in the 1930s, had been charged with looking at the linkages between minerals in soils and the nutrition of plants, animals and humans in the United States' (Bouis, 1995a, p12). Welch belonged to a network of scientists, linking PSNL with agriculture and food scientists at the university, promoting a 'food systems approach' to the problem of micronutrient malnutrition (Combs et al, 1996). This was envisaged as a holistic and 'inherently interdisciplinary' approach in which plant breeding research was integrated with research into micronutrient bioavailability[6] and food technology.

Welch introduced Bouis to Robin Graham from the Waite Agricultural Research Institute at the University of Adelaide.[7] At that time, Graham was conducting research 'to improve *plant nutrition*' by breeding for crops that have improved efficiency in the uptake of trace minerals such as iron and zinc from 'deficient soils, and which load high amounts of these minerals into plant seeds' (Bouis, 1995a, pp12–13, original emphasis). Graham's work had been concerned with improving wheat yields in zinc-deficient soils in Australia by developing varieties with higher zinc efficiency. At this point Bouis and his colleagues became aware of an existing NATO-funded programme to adapt Australian zinc-efficient varieties for the zinc-deficient soils in Turkey, with the aim of simultaneously improving both plant and human nutrition (Cakmak, 1996). It was estimated that 'Turkish wheat farmers would save $100 million *annually* in reduced seeding rates alone' (Cakmak, 1996, p13, original emphasis).[8]

As discussions had progressed between Graham, Welch and Bouis they envisioned a 'win–win situation' (Graham and Welch, 1996, pp15–16; Maitland, 1995; Cribb, 1995) promising the 'potential to enhance crops yields without additional farmer inputs and to improve their nutritional quality simultaneously' (Welch and Graham, 2004, p356). At this point, however, the linkages between the demonstrated agronomic effects and the potential nutritional benefits were still uncertain: 'Zinc-efficient genotypes absorb more zinc from deficient soils, produce more dry matter and more grain yield, *but do not necessarily have the highest zinc concentrations in the grain.* Although high grain zinc concentration also appears to be under genetic control, it is not tightly linked to agronomic zinc efficiency traits and may have to be selected for independently' (Bouis, 1995b, p18, emphasis added).

In 1994 IFPRI hosted an organizational workshop entitled 'Agricultural Strategies for Micronutrients' where these and other findings and perspectives were presented. The key outputs of this workshop were three working papers, generated to guide subsequent CGIAR research into biofortification (Behrman, 1995; Calloway, 1995; Graham and Welch, 1996). Of these, the paper by Graham and Welch has been the most influential. In summarizing this paper in an article for *SCN News* in 1995, Bouis stressed the key technical argument underpinning an initial phase of biofortification research, that of 'tailoring the plant to fit the soil' to generate the complementary agronomic and nutritional benefits: 'It is logical, then, to concentrate some breeding efforts towards breeding micronutrient efficient varieties for minerals that are required in small amounts, for which soil availability is low, but for which there are large reserves in the soil' (Bouis, 1995b, p18).

In this case, breeding for iron and zinc density was to be given a higher priority than vitamin A, which did not present the same win–win opportunity. As Bouis explained: 'We are starting our research on iron and zinc because they provide benefits to human nutrition and also have advantages for farmers. We would also like to research the viability of breeding plants for high vitamin A context. However vitamin A does not aid in plant nutrition and therefore will not provide higher yields' (IFPRI, 1995, p21).

It's in the seed

The various documents produced around this time, in particular the paper by Graham and Welch, convey a delicate balance. On the one hand, there is the belief in the potential of an idea:

> Two particular advantages accrue to micronutrient-efficient varieties if by virtue of their efficiency they accumulate more of the limiting nutrient in the grain. First, they offer better nutrition if consumed ... Second, such varieties have markedly better seedling vigour when resown on deficient soils ... Here there appears to be a win–win situation: the agronomic and human nutritional requirements coincide and reinforce the breeding strategy. An efficient cultivar with high seed micronutrient density will thus drive market forces simultaneously from both the consumer's and the producer's perspectives. (Graham and Welch, 1996, pp15–16)

On the other hand, there is an acknowledgement of an array of uncertainties about which there remained much to learn, in particular about human nutrition:

> Estimating the human dietary requirements of iron, zinc and vitamin A is beset with uncertainties. Homeostatic mechanisms are known to regulate the bodily absorption and excretion of micronutrients. Such mechanisms can maintain an individual in micronutrient balance even though daily dietary intake may be less than recommended safe for a healthy balanced diet ... The

amount of iron and zinc absorbed and used by a person can vary from less than one percent to more than 50 percent of the amount consumed and depends on a number of factors, including the types of food eaten, nutritional status, body demands and general health. (Graham and Welch, 1996, p4)

The merits of various breeding strategies were explored and debated in the paper, including breeding for increased micronutrient density in (or translocation from other parts of the plant to) the grain; increased micronutrient bioavailability 'promoters'; and/or decreased 'inhibitors' or 'anti-nutrients'.[9] In the process complex mechanisms regulating plant and human nutrition were discussed. In the conclusions, however, a crucial simplification is advanced, one which was critical to convincing CGIAR plant breeders to participate:

> From the exploration conducted to date, adequate genetic variation appears to exist to enable breeding of cultivars of the major food crops for higher micronutrient density ... The genetics of these traits is generally simple, making the task for breeders comparatively easy ... Where soils are deficient in one or more micronutrients, such high-density and high-efficiency genotypes will have a yield advantage ... *The primary selection criterion is a simple and efficient one – the micronutrient content of the seed.* (Graham and Welch, 1996, p55, emphasis added)

This framing of the problem spoke directly to the genetics-led approach to crop research within the CGIAR, allowing the conceptualization of grain micronutrient content as an 'isolable problem' (Anderson et al, 1991), which could be approached in same way as yield. However, uncertainties regarding linkages between mineral efficiency and grain content remained (Bouis, 1995b, p18). At this point, however, the language of holistic food systems was replaced by a more familiar CGIAR narrative – the answer was 'in the seed'.

A one-time, low-cost investment

Following the IFPRI-hosted organizational workshop in 1994, Bouis developed a proposal for a 'CGIAR micronutrients project'. A 'two stage process' was envisaged, reflecting a traditional CGIAR model of upstream research in which international public goods (IPGs) generated within international breeding centres cascade down through national adaptive breeding programmes to farmers in a variety of locations:

> The first, five-year phase will involve research primarily (but not exclusively) at CIAT, IRRI, CIMMYT, IRRI, PSNL and Waite. The cost has been estimated at approximately $2m per year for research on five crops [rice, wheat, maize, phaseolus bean and cassava]. During this phase, promising germplasm will be

identified and general breeding techniques will be developed for adapting nutrient-rich, high yielding varieties produced at these international agricultural research centres to specific growing environments in developing countries. During the second phase, the locus of the research will shift to national agricultural research centres and the focus will shift to adaptive breeding ... Certainly the annual costs for an individual country should not be more than the annual costs incurred by the five core agricultural research centres during phase one ... Thus the projected costs of a plant breeding strategy are relatively low as compared with the costs of supplementation ... The major part of the cost is the initial one-time cost of development. (Bouis, 1995a, p15)

Press releases at that time announced the prospect of 'nutrient-enriched crops ready for commercial production in 6 to 10 years' (Cribb, 1995, p44). At the same time, the investment required to carry out the research was compared favourably with the costs of distributing 'artificial' supplements: 'Bouis also argues that the $8 million projected costs of the research through the next five years runs far less than the $50 million it takes to provide mineral supplements to 28 million anaemic women in India' (paragraph printed in *News Republic*, 1995; *News-Sun*, 1995; *Observer-Dispatch*, 1995; *Times-Union*, 1995).

Therefore a win–win–*win* proposition was presented and the technical argument was combined with an economic one: 'A one-time investment in breeding research can offer a low-cost approach to fighting malnutrition that could improve farm yields at the same time' (Hilchey, 1995).[10] Bouis presented these arguments to several donors, but only the Danish International Development Agency (DANIDA) was prepared to commit funds at this point.[11] The project proceeded with modest funds; $1 million spread over 3 years, across five crops. The next section turns to activities around one of these crops, rice.

IR68144: 'A Serendipitous Discovery'

In the mid-1990s, a group of plant breeders at IRRI led by Dharmawansa Senadhira was working on 'problem soils' – salinity, mineral toxicity and deficiency. Senadhira had worked as a rice breeder for many years, from the late 1960s onwards, in his home country of Sri Lanka,[12] before joining IRRI in 1984. In 1996 he was appointed leader of IRRI's 'Flood-prone Rice Research Programme' and continued to work in close cooperation with NARS scientists to develop 'improved germplasm for flood-prone ricelands and for irrigated ricelands affected by low temperature and salinity' (Khush, 1998).

One of Senadhira's projects involved developing aromatic varieties for 'cold elevated areas'. Several materials were planted and one of these, a cross between IR72 and Zawa Bonday (a cold-tolerant variety from India), produced an elite line labelled 'IR68144'. Around this time, Senadhira heard about the micronutrients project, probably through contact with Howdy Bouis or Robin Graham.[13] While accounts of the precise timing vary, at some point the emphasis shifted to nutritional breeding objectives:

> So initially we tried to screen for the aromaticity ... It's one of the breeding criteria. And luckily, this IR68144 gave a good aroma. Because at that time, we submit this to the grain quality laboratory, they just rate it for strong aroma, moderate aroma, no aroma. So it's a bit rough, no analysis of zinc, iron, other [nutrients] in the grain ... Then, as the year goes by, Dr Senadhira tried to put this as a candidate for this micronutrient study ... So, from then on, simultaneously we have the germplasm, and working on the concept of the micronutrient project.[14]

These developments within the CGIAR were not alone in influencing this new direction. Filipino plant breeders in the group, in particular Senadhira's senior assistant, Glen Gregorio, saw the potential to address a serious nutritional problem in the Philippines: 'This research was influenced by the Philippine Government's efforts to eliminate the iron malnutrition problem in the country by artificially enriching consumption rice with iron' (Gregorio et al, 2000, p382; Graham et al, 1999; see the 'Iron Rice: The Silver Bullet?' section immediately below for further discussion of these efforts). In 1995 IRRI began collaborating with the University of Adelaide, where they were able to conduct 'mineral analysis according to international standards' (Gregorio et al, 2000, pp382–3). By 1999, 7000 samples, including IR68144, had been analysed for iron and zinc content (Gregorio et al, 2000).

In 1999, a seminar was convened by IRRI entitled 'Improving Human Nutrition through Agriculture'. This seminar brought together crop scientists and nutritionists to debate the findings of the CGIAR micronutrients project (1994–9), and was documented in a special issue of the *Food and Nutrition Bulletin* (vol 21, no 2, 2000). Presentations included updates on plant breeding, bioavailability and biotechnology research, and micronutrient programming experience. Among these were the results of the breeding work on trace minerals in rice by Senadhira and his colleagues, including the discovery and performance of IR68144 (Gregorio et al, 2000), which seemed to embody the 'win–win' argument for biofortification:

> A high-iron trait can be combined with high-yielding traits. *This has already been demonstrated by the serendipitous discovery* in the IRRI testing programme of an aromatic variety – a cross between a high yielding variety (IRT72) and a tall, traditional variety (Zawa Bonday) from India – from which IRRI identified an improved line (IR68144-3B-2-2-3)[15] with a high concentration of grain iron (about 21ppm [parts per million, sometimes expressed as μg/g] in brown rice). This elite line has good tolerance of rice tungro virus and excellent grain qualities. The

yields are about 10% below those of IR72, but in partial compensation, maturity is earlier. This variety has good tolerance to soils deficient in minerals such as phosphorus, zinc and iron. It has no seed dormancy and excellent seeding vigour, suggesting it could be a good direct-seeded rice. (Gregorio et al, 2000, p383, emphasis added)

Figure 2.1 *Field planted with IR68144-3B-2-2-3, IRRI, Los Baños*

Copyright: Michael Rubinstein, IFPRI, 2003 (reprinted with permission).

The 'serendipitous discovery' of IR68144 appeared to confirm a win–win relationship between nutrition and yield. The interpretation of this discovery as proof of this complex set of causal relationships served to black-box outstanding questions around interactions between plant nutrition, grain mineral content and crop yields in different environments. With these questions apparently closed, researchers now turned to the 'final key unknown': bioavailability (Haas et al, 2000, p440). While initial bioavailability studies had been conducted as part of the varietal screening process, using rats,[16] it was recognized that the next step was to research bioavailability in humans (Welch et al, 2000).

Iron Rice: The Silver Bullet?

As noted by Gregorio and his colleagues (Gregorio et al, 2000, p383) in the 1990s the Philippines government had already turned its attention to iron deficiency and to the possibility of fortifying rice with iron. Why this focus on iron rice? National prevalence statistics for iron deficiency anaemia produced by the Philippine Food and Nutrition Research Institute (FNRI) continued to indicate a significant public health problem: three out of ten children (including 66 per cent of infants aged between six and eleven months), four out of ten pregnant women and four out of ten lactating women are anaemic (FNRI, 2003, p71). At the same time, Filipinos rely on rice to provide much of their iron intake. Nationally aggregated statistics indicate that rice and rice-related products account for 34.2 per cent of total food intake (FNRI, 2003, p7) and 28.8 per cent of the total iron intake (FNRI, 2003, p10).

These statistics suggest that even a small increase in the iron content of rice could have a significant impact on iron deficiency at the national level. Which begs the question, why is such an initiative relatively recent? There is a long history of staple fortification in the Philippines, starting with an initiative to fortify rice with vitamin B (to address beriberi which was then a serious health problem), which commenced in the mid-1940s and was formalized in the Rice Enrichment Law of 1952.[17] At that time, resistance from the rice millers, who saw the law as a means to monitor and tax their income, presented the main obstacle to implementation. The highly decentralized nature of rice milling (currently there are more than 10,000 registered rice millers in the Philippines, compared to 6 flour mills) therefore made this law difficult to enforce. In time beriberi ceased to be a public health problem and the law was increasingly ignored, though never rescinded.[18]

The Nutrition Act of the Philippines was passed in 1974, through a presidential decree (the Philippines was then under martial law), which established nutrition as a national priority. This Act remains the framework for nutrition programming in the Philippines and has been the basis for six successive medium-term national nutrition plans (Florencio, 2004, p5). Throughout the 1980s and 1990s there was a steady expansion of nutrition programming, notably fortification of a variety of staple foods, including wheat flour (iron), oil and margarine (vitamin A), refined sugar (vitamin A) and salt (iodine), often through public-private partnerships (Solon, 2000). Fortification of these food items has been less problematic than rice, given their more homogeneous nature and the centralized organization of their production and processing. More recently, an increasing range of fortified items, notably instant noodles and carbonated drinks, is becoming available in urban areas. This has led critics to suggest that 'unbridled fortification' is giving foods that are otherwise 'empty calories' an 'aura of being nutritious' (Florencio, 2004, p58).[19]

However, such interventions represent a tinkering around the edges of what remains an overwhelmingly rice-based diet (Balgos, 2005). In the

Philippine context, therefore, high-iron rice represents the ultimate magic bullet. In contrast to other staples, however, rice markets are highly differentiated and subject to strong consumer preferences.[20] Furthermore, as I was constantly reminded, in the Philippines, 'rice is a very political crop'.[21] In a culture in which 'rice is our daily bread' (Castillo, 2006, p80),[22] any fluctuation in its price or availability is likely to be politically sensitive. In this context, attempts to add value to rice through chemical fortification raise difficult questions about who will bear the cost, particularly if the target group for such an intervention is poor consumers.

In the early 1990s, the Fidel Ramos administration turned its attention to iron rice fortification, which due to these logistical and political conundrums had never attracted the interest of the private sector. In this case the FNRI, as the governmental nutrition research body, was charged with conducting the research. In fact, FNRI researchers had been developing an iron rice mix since the 1980s on the strength of their survey data, 'but there was no push yet at that time'.[23] However, as noted earlier in this chapter, in the early 1990s there had been a renewed interest at the national level following the 'Ending Hidden Hunger' international conference. In response to this, FNRI researchers developed methods and fortificants, and conducted a series of bioavailability and efficacy studies.[24]

The Food Fortification Act

The accumulated experience of an active nutrition community enabled them to lobby for a more comprehensive legal framework for food fortification. The Food Fortification Act (Republican Act 8976) was passed in 2000 and came into force in 2004. Under this Act, fortification of the main food items in the Filipino diet, according to FNRI surveys, was made mandatory: rice with iron; wheat flour with vitamin A and iron; refined sugar with vitamin A; and cooking oil with vitamin A. In addition, food producers are encouraged to voluntarily fortify other foods through the 'Sangkap Pinoy Seal Program'.[25]

In the case of mandatory fortification, responsibility for each food item was allocated to a government agency, with rice fortification delegated to the National Food Authority (NFA, formerly the National Grains Authority). Initial implementation was directly though supplies of NFA rice, which accounted for 10 per cent of the total rice market in the Philippines and which, priced at Php18/kg, is usually the cheapest rice available in urban markets. Since NFA rice was milled centrally, this avoided the problems associated with decentralized milling that were encountered in the earlier case of vitamin B-enriched rice. Furthermore, dilemmas about how the additional costs of fortification would be distributed were also shelved, at least temporarily, as the NFA absorbed the costs of fortifying its own supplies. At this point, enforcement of fortification throughout the commercial market was a long way off, with fortification of NFA rice postponed from 2004 to graduated targets of 20 per cent by 2007, 40 per cent by 2008 and 100 per cent by 2011.[26]

The more immediate obstacle, however, was the limited supply of the chemical fortificant. While research continued at FNRI, the iron pre-mix was procured from a US-based private company.[27] Since NFA rice was imported from Vietnam, the arrangement was to transport the pre-mix to Vietnam, fortify with the pre-mix *in situ* (where labour costs are lower) and import the iron-fortified rice (IFR) into the Philippines. At that point the IFR initiative started to become part of the more contentious politics of rice imports, a sensitive issue in the Philippines (Dano and Obanil, 2005; Castillo, 2006; Dawe et al, 2006). However, the NFA was compelled to continue with this arrangement so that it was able, at the very least, to supply sufficient IFR supplies for the Hunger Mitigation Program of the Office of the President (Gloria Macapagal-Arroyo), in particular the highly politicized 'Food for School' programme in selected schools in the Metro Manila District. It was at this point that NFA officials began to see potential for iron biofortified rice as offering a way out of these political and logistical dilemmas.[28]

Proof of Concept: The Sisters of Nutrition

Dr Angelita Del Mundo, a nutritional anthropologist at the Institute of Human Nutrition and Food (IHNF) in the Centre for Human Ecology at University of the Philippines, Los Baños (UPLB), had been working on quality assessment of rice since the 1950s. Situated on the Los Baños campus, close to IRRI and the University's Agronomy Department, Del Mundo had, for many years, been 'a missionary for integrating nutrition into rice'.[29] In 1996 Bouis approached her as a possible research partner to work on rice biofortification. While UPLB plant breeders were sceptical, Del Mundo decided to proceed, arguing that 'if we don't test the waters: who will?' Moreover, Del Mundo 'had always dreamed of working with religious sisters' in her research work (Del Mundo, 2003, p82). These elements came together in a study considered the first of its kind (Haas et al, 2005) to test the efficacy of iron-biofortified rice with religious sisters as 'actor-subjects'.[30]

Del Mundo and her assistants, Angelina (Ning) Felix and Melanie Narciso, began with family studies. However, they found it was difficult to measure nutritional impact in such a small-scale and uncontrolled setting. Meanwhile, they were developing the idea of conducting their research in Catholic convents, which they felt could provide a more structured setting to study the bioefficacy of iron-biofortified rice. In 1999 they conducted a pilot study in a convent, with 27 sisters. Results indicated some improvements in iron status, but this could not be attributed to the consumption of IR68144 in the absence of a control group. However, this experience convinced Del Mundo of the suitability of the convent as 'an ideal research setting': 'In summary, the high prevalence of iron deficiency, the considerable amount of rice consumed, the excellent cooperation of the subjects, and the structured routine of the convent make this an ideal setting to investigate the effect of improving iron intakes through a staple food' (Haas et al, 2000, p442).

Del Mundo presented the findings of her pilot convent study at the 'Improving Human Nutrition through Agriculture' conference at IRRI in 1999. On the strengths of these initial findings, she outlined a proposal to conduct, with collaborators Jere Haas of the Division of Nutritional Sciences at Cornell and John Beard from the Department of Nutrition at Pennsylvania State University, a large-scale, controlled, double-blind study in ten convents in and around Metro Manila to test the bioefficacy of IR68144 (Haas et al, 2000, p442).

One of the participants at the IRRI seminar was Joseph Hunt of ADB. Hunt agreed to fund the rice component on the condition that the feeding trial – as the bioefficacy study was popularly known – be included in the project and additional funds were secured for this purpose.[31] In addition, funding was provided to support the position of a CGIAR micronutrient project director, enabling Bouis to focus on the project on a full-time basis for the first time. The rice biofortification component, now entitled 'Rice Breeding to Reduce Anaemia in Asia (2001–3)' with participating countries including the Philippines, Bangladesh, Indonesia and Vietnam, was incorporated into a broader programme 'Improving Nutrition of Poor Women and Children in Asia'.[32]

Bioavailability: The final unknown?

It is interesting that such momentum gathered around the proposal for a bioefficacy study, given that the starting point was an acknowledgement that bioavailability represented the critical knowledge gap. A typical model for assessing efficacy of nutrient-enriched foods for human nutrition usually includes the following stages:

1   Assessment of nutrient bioavailability in humans (possibly preceded by animal *in vivo* and/or *in vitro* studies);
2   Assessment of efficacy for improving human nutritional status (bioefficacy);
3   Assessment of effectiveness at the community level (incorporating acceptance and regulatory factors) (King, 2002).

In this case, however, the bioavailability stage was bypassed and the first human studies were for efficacy, not bioavailability. Nevertheless, an expectation was created – and subsequently sustained – particularly among crop scientists, that such a study would simultaneously answer questions about bioavailability. This begs the question: why was a human bioavailability study not conducted first? In contrast to a nine-month bioefficacy study, a human bioavailability study would have taken two to three weeks.[33] For researchers who were at the centre of the study at that time, the answer was clear: 'the objective was to attract donors', so 'proof of concept was critical'. In this case a bioefficacy study (rather than bioavailability research) would provide the required proof of concept by demonstrating 'a biologically significant effect'.[34]

Constructing the study

The iron rice project brought together the two strands of biofortification research: plant breeding and human nutrition. While initially identified for its elevated iron and zinc content, as well as certain agronomic characteristics, IR68144 now took on the identity of 'high-iron rice'. This identity, however, had two facets: as experimental material used in a controlled study and as germplasm with commercial potential. This section deals with the role of IR68144 in the feeding trial; the next discusses its assessment and release as a high-iron variety. As will become clear, over time these two processes became increasingly conflated.

Source: Angelina Felix, Institute of Human Nutrition and Food, University of the Philippines Los Baños (reprinted with permission).

Figure 2.2 *Sisters dining: Bioefficacy study participants and researchers at one of ten participating convents in Metro Manila*

After the IRRI seminar, the emphasis shifted to nutrition research, with the high-iron IR68144 as the material 'developed at the IRRI for experimental use' (Haas et al, 2005, p2825) in a large-scale, controlled bioefficacy study across ten convents. As already mentioned, this was preceded by smaller scale and pilot studies, which led the researchers to conclude that convents would provide a suitably structured research environment in which subjects would be representative of the proposed target group for high-iron rice. The study design has been described as follows:

This study was a prospective, randomized, controlled, double blind, longitudinal (9 mo) intervention trial involving 317 women. The study had 2 arms: low-iron rice and high-iron rice, which were the exclusive sources of rice consumed for 9 mo. Randomization was done according to 2 strata based on ferritin (Ft) and Hb concentrations. At each of 10 research sites women, who had a Hb concentration < 120 g/L or a Ft concentration < 20 µg/L at baseline, were randomly allocated to 1 of 4 rice groups, followed by randomization of all remaining women at the site. For the duration of the study, all of the participants, as well as the field workers, were unaware of the subjects' rice group assignments. Each rice type was randomly assigned 2 different colors in each convent. Colors used were green, blue, gray, and cream. Rice was delivered in bins of these colors, cooked in containers with these matching colors, and served in bowls of the corresponding color. (Haas et al, 2005, p2824)

In preparation for this study, researchers at IRRI conducted a series of experiments to measure the effects of varying factors in the production, milling and cooking of IR68144 (the proposed 'high-iron rice'), in comparison with the proposed control rice variety, PSBRc28 (the proposed 'low iron rice'). While this study provided new information on the effects of environmental and seasonal factors on mineral content and yield, the key finding for the purposes of the bioefficacy study was that: '*Post harvest practices remain a critical factor in enabling the grain to retain its Fe content*' (Gregorio et al, 2003, emphasis added).

Furthermore, the researchers concluded that, subsequent to milling, washing and cooking, the differential between IR68144 and PSBRc28 in terms of iron content would be insufficient to demonstrate nutritional impact in the planned feeding trial:

To test the bioefficacy of Fe of IR68144, a 5.5ppm differential between IR68144 and its control is necessary. Initially, PSBRc28 was chosen as the control rice to IR68144 but this study showed that the PSBRc28 as having a comparable amount of Fe to IR68144 [sic]. The difference in Fe content may only be achieved by applying different milling degrees to the two varieties. The differential in favor of IR68144 is poorly and inconsistently expressed in the data from large scale planting. The reason is not clear but location/ soil/ environment effects may not have been random or equal and may have worked against IR68144. *In effect, the differential between IR68144 and PSBRc28 is largely based on milling and not genotype* ... It was also noted that different cultural practices were applied in planting IR68144. The Differential may be achieved if a commercially produced rice such as C4 will be used opposite to IR68144 for bioavailability studies but treatments such as milling of IR68144 and washing of rice prior to cooking should be taken into consideration to maximize the differential [sic]. (Gregorio et al, 2003, emphasis added)

These conclusions highlight a number of factors which accumulated to affect the iron content of iron-biofortified rice as it is eaten (aside from additional factors introduced when rice is part of a meal): environmental factors and cultural practices in rice planting; milling method and degree; and food preparation practices. Notwithstanding the challenges this might present to commercializing the variety, these findings had a significant effect on the study design. To ensure the required iron differential was maintained, the high-iron rice was undermilled and the control rice, now C4, overmilled. Since undermilled rice does not store well, the high-iron rice was milled and delivered to the convents on a fortnightly basis.[35] The preparation and cooking of rice (only rice, not other food items) was standardized by assignment of this task to the field assistants based in each convent. Rice portions were controlled through weighing and samples of cooked rice were taken for regular analysis to monitor iron content (Haas et al, 2005, p2825).

A research family

While the account so far describes some of the technical challenges in establishing what was an ambitious study, and how these were overcome, it does not take account of the social 'glue' that held this interdisciplinary, multi-institutional research effort together. The Filipino researchers at IRRI and IHNF who coordinated the research and communication of findings frequently refer to this group as 'the family', which included, in addition to Del Mundo, Gregorio, Felix, Narciso, Cristina Sison[36] and Dante Adorada at IRRI. Also included in the family, as 'the grandfathers', were Bouis, Welch and Graham. Even the more recent US-based recruits to the project, Haas and Beard (affectionately known as 'the iron man'), seem to have been included in this conception of a research team as an extended family. Furthermore, an important contribution to the 'glue' holding this extended family together was no doubt a sense of respect for the late mentors, Senadhira and (in the later reporting stages) Del Mundo,[37] who was identified most closely with a shared vision of the project.

The way in which this group of Filipino researchers saw their research network as a family reflects a particular aspect of Filipino culture, represented by the term *barkada*. As the following blog introduction indicates, the *barkada* is a flexible concept for describing social relations and networks characterized by 'closeness': 'In Filipino, the word "barkada" means a group of friends. As with many things Filipino, the delineation of closeness is not exact. I've heard it described as a group of close-knit friends or simply a peer group. The best definition I can come up with is that to a Filipino, one's barkada is another form of family.'[38]

What were the ties that bound this *barkada* together? Three factors stand out. Despite its international membership, the family home was in the Philippines.[39] This reflected not only the immediate Filipino membership, but also Bouis, considered Filipino one-step-removed due to his long-term professional and personal relationship with a country to which he first travelled, many years earlier, to carry out his PhD field research. The second factor was the focus on iron,[40] addressing a nutritional problem whose status as a national priority was undisputed (in contrast to the case of vitamin A, as will be seen in the next chapter). Finally, the centrality of religious sisters and the institutional setting of the convent seemed to highlight the integration of humanitarian and religious elements within a single project.[41] This was further emphasized by the notion that Del Mundo and Bouis were akin to 'missionaries'[42] in their determination to make iron-rich rice a reality.

Furthermore, this study was framed as a 'food systems based approach' as defined by Del Mundo and her colleagues at the IHNF. In this case, the aim was not to meet daily requirements with one food item such as rice, which would constitute a 'curative' strategy, but to complement other foods and types of intervention with a 'preventative' strategy. This strategy was based on a notion of biofortification as 'nutritional enhancement ... enhancing what is *there*'. Moreover, as mentioned earlier, the project attracted such a level of commitment '*because* it was iron'.[43] In this case, therefore, notions of 'impact' were built on the understanding that even a modest improvement in the iron content of rice was a valuable, incremental step, one of many strategies employed towards the achievement of a larger goal.

With these aims in mind, the IHNF researchers had approached selected convents to negotiate access for the purposes of the feeding trial. This was not an easy task as religious congregations are not accustomed to allowing researchers to visit, let alone stay in the convents for such a lengthy period. Some researchers, themselves Catholics, felt uneasy about entering convents, at least initially, on a pretext of feigned interest in joining the congregation. In each case, gaining the trust of convent elders was difficult yet crucial. At one point, a local newspaper headline which read 'Nuns as Guinea Pigs' almost ended the pilot convent study, and with it the chance of conducting the larger study that was planned.[44] As the IRRI seminar was in process, Del Mundo and Felix missed one day in order to visit the convent and assure the elders that this article had misrepresented the researchers' intentions. Finally, the leadership of ten convents agreed to allow the researchers to conduct the study. In retrospect, the turning point seems to have been when elders were convinced that their participation would constitute a form of 'humanitarian service'.[45]

It has been suggested that this study brought together science and religion,[46] a statement that many expatriate scientists find puzzling. A catchy headline chosen for the launch press release, 'The Sisters of Nutrition' (IRRI, 2000a), continues to generate mild amusement. At the same time, questions have been asked about the reality of 'prior informed consent' in such a strictly enforced, hierarchical setting.[47] However, discussions with the Filipino researchers at the centre of the study reveal a particular character of research relations that had unusual reciprocal and spiritual dimensions. For example, in December 2006, I accompanied the researchers as they made their annual pre-Christmas visit to the participating convents to deliver rice (high-iron rice) as 'thanksgiving to the sisters'. One of the researchers at the time reflected that 'maybe because I am Catholic ... they are part of me too', and recalled instances, in the busy schedule of the study, when 'they would say to me "you look tired, but *go on*, you are doing good work"'.[48]

From IR68144 to MS13: 'A Special Variety'

As the researchers were preparing to start the bioefficacy study, IR68144 was submitted to the National Cooperative Testing (NCT) programme of the Rice Varietal Improvement Group (RVIG) in the Philippines. This is the national programme which assesses all germplasm for release in the Philippines, whether generated by IRRI, PhilRice or a private company. These institutions normally submit germplasm to this testing programme having first conducted their own station trials. Still, promising potential varieties can fail, since the programme involves multi-locational tests over three years, and consistent performance over the entire period is required for certification.[49]

According to the researchers, IR68144 was 'visualized as for national release'[50] from the start. Developing new germplasm is a lengthy process, taking eight to ten years. For plant breeders, a new variety is never the final product, but one of a succession of improved models 'like Honda cars'.[51] So it is not surprising that, given that the NCT phase takes three years, IR68144 was submitted at the beginning of the feeding trials on the assumption that the publicity accompanying positive results from the feeding trial might help promote it as a newly released variety. However, IR68144 presented a dilemma to the RVIG since there are no existing criteria or guidelines for assessing nutritional traits. Two options were available: as a certified variety with the National Seed Industry Council (NSIC, formerly the Philippine Seed Board) certification, or as a 'Maligaya Special'[52] or MS variety. NSIC varieties are 'high-yielding' varieties; MS varieties do not have to meet stringent yield parameters but are instead assessed for particular desirable characteristics and include aromatic, pigmented and glutinous rice varieties.

Since the yield performance of IR68144 did not meet the agronomic performance requirements for an NSIC variety, it was put forward for assessment as a special variety. Its 'slight aroma' seems to have been one factor in this decision. However, it was accepted for assessment within this category primarily on the basis of its nutritional characteristics. And, since the RVIG did not have the guidelines or means to assess these traits, the results provided by IRRI were accepted at face value.[53] Interestingly, the data available at this point was IRRI's initial assessment of iron content as 21ppm (Gregorio et al, 2000, p383), a measure that was revised downwards once the bioefficacy study was under way and further compromised as a result of better understanding of the impact of post-harvest losses, particularly during milling (Gregorio et al, 2003).

IR68144 passed the assessment and was released as a special variety, named MS13,[54] in November 2003. The following report indicates a rather ambiguous final evaluation, which stands in contrast to the optimistic assessment outlined earlier by Gregorio et al (2000, p383), reminding us again of the role of serendipity in its selection:

> IR68144-2B-2-2-3-2 is an aromatic line serendipitously discovered to contain high grain concentration. Based on the field performance tests, this line gave a modest yield in spite of its susceptibility to insects, pests and diseases. Overall performance indicated its yield potential in the five-ton category, where most of the PSC or NSIC varieties fall. Its early maturity and tolerance to tungro disease and enhanced iron content are value added traits that could possibly attract the adoption of this line. Its novelty was appreciated by the members of the RTWG[55] and was recommended to be named as MS13 in the category of Maligaya Special rices. In contrast to protein which has negative effect on yield, the relationship of yield and micronutrient-dense rice maybe [sic] positive in some minerals ... *It may not be a truly impressive performance but its discovery catalysed the inclusion of nutrition as one of the breeding objectives.* (Padolina et al, 2003, p11, emphasis added; see also Corpuz-Arocena et al, 2004).

Contested findings

By the end of the ADB project, IRRI plant breeders were quoting iron levels of polished IR68144 as 4–5ppm (in contrast to the initial 21ppm), which equates to double the iron content in commercially available varieties.[56] This more conservative figure reflected expected post-harvest losses in normal commercial conditions and variability of measured iron content of IR68144 when planted in different environments. At the same time researchers involved in the feeding study maintained that the differential was in the region of 400–500 per cent, referring to empirical evidence from the repeated measurement of cooked samples of IR68144 and the C4 control used for the feeding trial (Haas et al, 2005, p2823).

These contradictory viewpoints generated a number of debates, reopening the black box that had previously closed around grain mineral content. Clearly it was not simply a matter of how much iron was in the grain, but *where* in the grain. The preparatory studies conducted at IRRI (Gregorio et al, 2003) indicated that most of the iron content was contained in the aleurone layer normally removed during polishing, so that 'the differential between IR68144 and PSBRc28 [was] largely based on milling and not genotype' (Gregorio et al,

2003). Given these new insights into the distribution of iron within the grain, would a shift downstream from plant breeding to post-harvest research prove a more productive line of enquiry?

The GxE (genotype by environment)[57] findings from multi-locational trials generated a yet more fundamental question about the initial premise of biofortification (through conventional plant breeding); that sufficient genetic variation existed (Graham and Welch, 1996; Bouis et al, 1999) and could be accessed through the 'simple and efficient' screening criterion of 'the micronutrient content of the seed' (Graham and Welch, 1996, p55). Reflecting on the strong influence of environmental factors, scientists specializing in soil science and cereal chemistry have pointed to the potential for studies of environmental factors and cultural practices optimizing iron expression to inform the initial choice of breeding parents. In so doing, they offer perspectives reminiscent of an earlier conceptualization of biofortification as 'tailoring the plant to fit the soil' (Bouis, 1995b, p18), highlighting the difficulties faced by national programmes tasked with seed production:

> The basic problem ... there is not enough information on what will optimize the expression of iron ... They identify genotypes first, without understanding the cultural practices that will optimize the expression of iron. [Then] national institutes start breeding, but they don't know what practices to use ... *If they did an agronomic study first, then the true breeding parents would have been used.*[58]

Observations such as these have led these scientists to question the assumptions about the predictability of GxE interactions on which biofortification, as a viable approach, rely:

> If the trait is genetic ... there shouldn't be such variability ... These environmental tests should have been done at the beginning ... I strongly feel there is a significant influence of environment ... breeders should not only be looking at the content of the grain ... look at the ability of the plant to absorb iron ... look at the root system ... the absorbing capacity of the plant.[59]

The above debates point to differences between epistemic cultures within the plant sciences. Notably, in the case of GxE interactions, members of the classic cluster of crop sciences are 'divided by a common language' – all talk about GxE, but mean very different things. For plant breeders, GxE is 'dealt with' through multi-locational trials, from which (provided results fall within a certain range) the mean result is taken. Soil scientists, on the other hand, interpret environmental variation as an entry point that raises new research questions.[60] It is this contrast between a fundamental orientation towards universality and replicability, on the one hand, and one towards context-responsive enquiry and specificity, on the other hand (Biggs and Clay, 1981), that has been key to securing the hegemonic position of plant breeding. In the context of an international system that has consistently sought the optimal and universal, a discipline that privileges the 'G' over the 'E' is able to deliver the required level of certainty.[61]

While previous discussion highlights enduring, asymmetric relations between the crop sciences, the case of rice biofortification project introduced another crucial interface: between the plant and human sciences and between the epistemic cultures of plant breeders and nutritionists. Until this point the bioavailability question had represented a challenge put by nutritionists to plant breeders promoting biofortification as a viable nutrition intervention. In this case, debates evolving at IRRI in the wake of the feeding trial represented a point of departure. Power relations between the two epistemic worlds were shifting and, in the process, the dynamic concept of bioavailability as a multiplicity of interactions was domesticated and reconstructed as a simple, fixed metric, amenable to plant breeding-led research.

As mentioned earlier, the feeding trial was a bioefficacy study that inadvertently carried with it an expectation that it would simultaneously provide data on bioavailability; preferably a 'bioavailability number'[62] that plant breeders could work with. While this was not provided, the study findings supported (but did not confirm) existing assumptions of a 10 per cent figure for iron bioavailability in rice: 'Although these theoretical calculations support an estimated 10% efficiency in the transfer of iron from rice to body stores, the actual bioavailability of the IR68144 rice variety after processing and in the context of a typical Philippine diet has not been determined' (Haas et al, 2005, p2829).

This qualified statement was subsequently reframed as a definitive finding: 'The study also indicated that iron bioavailability is 10% in cooked rice' (Gregorio and Haas, 2005, p39).[63] This suitably neat and memorable figure gradually became accepted within the IRRI community as the iron bioavailability number for rice in the typical Philippine diet, 'black-boxing' the unanswered questions still surrounding bioavailability and foreclosing future scientific enquiry into this issue.

At the outset, this study was seen as just the beginning and, as expected, it generated a range of new researchable questions around agronomic and post-harvest practices and nutritional factors. It was at this point, however, that the family began to disperse and questions such as these have yet to be identified or followed up. While some members had already moved on, with the reassignment of Gregorio (awarded 'Outstanding Young Scientist' in 2004)[64] to the African Rice Centre (WARDA) in January 2006, the network finally lost its critical mass and momentum. At this point new faces arrived at IRRI, and these issues and debates carried over into the newly established CGIAR-wide HarvestPlus programme, discussed in Chapter 4. Meanwhile, as the 'family' dispersed and debates about international bioavailability and GxE remained unresolved, closer to home the science and politics of rice was moving in a very different direction. The next section focuses on events taking place on the national stage at the time, which shifted attention away from the release of the first nutritional rice variety in the Philippines to more immediate food security concerns.

National Release: In the Shadow of Hybrid Rice

MS13, the high-iron 'special variety' released following national, multi-locational testing of IRRI's IR68144 germplasm, was issued in the Philippines in November 2003. At this point it would have been the role of PhilRice to promote the newly released variety, yet this did not happen. This may have been due to the shortcomings of the variety; its agronomic characteristics were mixed at best; it was only slightly aromatic; even its nutritional characteristics had been reassessed. An additional factor may have been the way the plant looked. Another 'serendipitous discovery' (Barker and Dawe, 2002, p4), IR-8, still a potent symbol of the Green Revolution despite its brief life in the fields, was 'a beautiful rice plant' (Castillo, 2006, p129). Reactions to MS13 could not have been more different: 'If you see the stand of the material you would not be convinced to plant it ... at harvest the leaves are dried up ... if you are a farmer you would not be convinced to plant the material.'[65]

However, to understand why MS13 was sidelined it is necessary to recognize other developments which took place in the Philippines in the years leading up to 2004, the 'International Year of Rice'. In particular, as the following sections illustrate, the Arroyo administration intensified its promotion of hybrid rice as the answer to the national priority of rice self-sufficiency, casting a shadow over other ongoing activities.[66]

Rice self-sufficiency is national security

While the prospect of iron rice appeared timely, given the passing of the Food Fortification Act (and the challenges faced in implementing an iron-fortified rice programme receiving presidential backing), this has been easily overshadowed by a programme aimed at one of the most central issues in national politics. The issue of rice self-sufficiency has been a high-profile feature of Philippine political life since before the Green Revolution, and the prospect of rice imports strikes at the heart of Philippine national identity and notions of national and food security:

> It has always been said that 'Rice self sufficiency is national security'. It is a matter of celebration when it is achieved and a matter of shame and blame if it is not. Needless to say, rice self-sufficiency has positive value just as rice shortage, with delayed importation and increase in rice prices can bring political misfortune to someone. Such was the case of the '1995 rice crisis', the principle cause of which was 'the failure of the government to anticipate a shortfall in domestic production and to plan imports

to make up to the shortfall.' Along with this record came the 'downfall' of the Secretary of the Department of Agriculture at that time. Because of the high political cost, no politician wants to get caught with a rice shortage, increased rice prices, and worst of all, queues of urban consumers waiting to buy cheap rice especially on a rainy day. (Castillo, 2006, pp33–4)

Such a national preoccupation with self-sufficiency may seem surprising, given the track record of the Philippine rice economy which since the 1950s has consistently shown net imports, with only brief period of self-sufficiency in the late 1970s (Castillo, 2006, p33). In 2006, the Philippines imported 10 per cent of its rice supply, compared to 30 per cent in Malaysia (Dawe et al, 2006, pxiv). Dawe et al contend that the political dimensions of the problem are overstated and suggest the main reason the Philippines is a net rice importer is its geography; as a nation of islands it cannot compete with countries on the Southeast Asian mainland such as Vietnam and Thailand, which benefit from large river deltas. Indonesia, the Philippines, Sri Lanka, Japan, Korea and Malaysia, all islands or narrow peninsulas, 'have been consistently importing rice for more than a hundred years' (Dawe et al, 2006, pix).

Nevertheless, with memories of the 1995 rice crisis and the prospect of international rice markets 'racing to fill China's rice bowl', it seemed likely that the government would remain in pursuit of 'rice self-sufficiency at any cost' (Castillo, 2006, p49).[67] In 2003, Philippine NGOs and social movements successfully lobbied the Arroyo government to abandon plans – developed with the encouragement of international agencies – to liberalize the rice market, and instead to retain quantitative restrictions (Berber, 2003, pp10–23): 'Along with South Korea, the Philippines remains one of only two countries in the WTO that maintain quantitative restrictions (QRs) on rice imports' (Tolentino, 2002, p156). With this avenue closed, the government refocused on the issue of rice yields and directed its attention on an initiative under way in the Philippines since 1998, the hybrid rice programme.

Focus on yield: 'Gloria Rice'

Hybrid rice technology was developed in China in the 1970s and is credited with producing dramatic productivity increases achieved during the period 1975–1990 (Obanil and Dano, 2005, p22). The Hybrid Rice Commercial-ization Program (HRCP) of the government of the Philippines was launched in 1998; however, initial adoption was slow.[68] In 2001, the government provided an additional incentive in the form of a subsidy scheme in which 'farmers pay only 50 percent of the price of the seed and the government pays for the remaining 50 percent' (Norton and Francisco, 2006, p158). In 2002, under Executive Order (EO) 76, PhilRice was transferred from the Department of Agriculture to the Office of the President in order 'to intensify the government's hybrid rice programme' (PhilRice, 2002, p20). Since then President Gloria Macapagal-Arroyo has personalized hybrid-rice promotion in a style that typifies the *palabas* or 'showiness' that permeates Philippine politics (Cullather, 2004, p236): labelling the government's flagship hybrid variety Mestizo as 'Gloria Rice' for a press briefing (Laguna, 2002, p8).

Hybrid rice had been controversial, however, for a number of reasons. The programme relied on seed production, not only by public institutions such as PhilRice, but also private companies including Bayer Crop Science, HyRice Corporation and SL Agritech (PhilRice, 2006, p18). While the intention was to nationalize hybrid seed production within the cooperatives, critics were sceptical about whether this could be achieved, particularly when the subsidy period ended.[69] Furthermore, the use of government funds to subsidize private-sector produced seeds was itself controversial. Second, unlike open pollinated varieties, farmers cannot save hybrid seeds from the previous season, so become 'locked in' to buying seeds each year. Third, from the perspective of NGOs and social movements promoting alternative visions of sustainable agriculture, the highly subsidized hybrid rice programme distracted attention and funding away from more sustainable solutions such as organic agriculture, which they argued was more suited to a large majority of Filipino rice farmers, who have small land holdings and are increasingly burdened with high input costs (Obanil and Dano, 2005).

From nutrition to hunger mitigation

When biofortification and high-iron rice are placed in the context of these developments, the reasons for the marginalization of MS13 become clearer. With its eye on imperatives of self-sufficiency and access, the main concerns of the government were towards increasing productivity and lowering prices. In this case, nutritional enhancement, whether by fortification or biofortification, was a subsidiary concern. Here, there was a striking contrast between the high-yielding varieties of the Green Revolution, which in their day struck a chord with the priorities of the Marcos government (Cullather, 2004)[70] and the dissonance between a powerful and enduring national self-sufficiency narrative and the complications and trade-offs presented by nutrient-enriched rice.

These dynamics illustrate how localized biological and socio-political factors combined to thwart the promotion of the product for which the iron rice family had held such high hopes. Despite their location in the Philippines, this nevertheless partial network had failed to predict the particular combination and timing of events that would overshadow the project that had been so central to their world. In the event, even an initiative based around MS13 in Laguna province – one of the locations in which the germplasm had performed relatively well in agronomic trials – and sponsored by the provincial governor (ANGAT-Laguna, 2006; *Southern Tagalog Herald*, 2006, pp1–2) was faltering, since farmers were unimpressed by the appearance of MS13 in the field, particularly when compared with hybrid rice, which has benefited from a generous government subsidy. As the coordinator exclaimed: 'How I wish I had the same subsidy [for iron rice]! ...

hybrid rice ... standing there, looking so stately ... then you see this lowly thing ... who will grow that?'[71]

Why, then, did the government commit to the iron-fortified rice (IFR) programme and incorporate it into its high-profile Food for School initiative, perhaps unwittingly drawing attention to a national dependence on rice imports? This is a difficult question to answer; government officials and nutritionists point to the evidence base underpinning the fortification law, as well as imperatives emerging from the MDG interim review in pushing the government to take action over micronutrient malnutrition. However, national-level discussions about nutrition policy are increasingly channelled towards the more emotive and narrow question of 'hunger mitigation', propelled by high-profile media coverage of hunger surveys conducted by groups such as Social Weather Stations (SWS).[72] In this case, food security – not nutrition – was the overriding concern. In this context the Food Fortification Act passed without opposition. While issues around rice and hunger were highly politically charged, nutrition remained 'politically neutral' for government and civil society alike.[73]

Conclusion

This chapter has traced the history of early research into the possibility of breeding high-iron rice varieties. This research brought together various disciplines including plant breeding, plant nutrition and human nutrition, disciplines that converged in the iron rice study in the Philippines. Was this process indicative of the emergence of a new, integrative science linking a CGIAR centre, IRRI, with national and international researchers? The initial conception of the research around the holistic concept of food systems certainly appeared to point in this direction.

This account, however, reveals how episodes of open-ended, holistic and systems-oriented enquiry were punctuated at critical moments by an institutional reflex to simplify and tailor unfamiliar concepts and black-box uncertainties in such a way as to maintain existing interdisciplinary divisions and hierarchies. In the process a genetics-led approach, long established within the design and culture of the CGIAR in general and IRRI in particular, progressively framed out agronomic and nutritional variables from an increasingly reductionist notion of grain iron content as 'an isolable problem' (Anderson et al, 1991, p32), which could be manipulated in a similar way to agronomic traits, such as yield and stress tolerance.

For a period of time, however, a close-knit interdisciplinary network formed around the feeding trial with the 'Sisters of Nutrition' and the potential of a high-iron rice variety, IR68144/MS13, to transform the possibilities of plant breeding for improved nutrition. This interdisciplinary 'family' succeeded in demonstrating the bioefficacy of iron in high-iron rice and in securing the national release of IR68144 as MS13, albeit as a 'special' variety. To date, MS13 is the only biofortified rice variety on the market; yet it has not been actively promoted and adoption has been very limited. Here the politics of hybrid rice and national food self-sufficiency collided with local socio-cultural dynamics, altering the course of events in ways the 'family' could not control. In this case, the politics and science of rice came together in the Philippine context, in a very particular way, in the year preceding the 'International Year of Rice'.

The interdisciplinary iron rice research 'family' was, in effect, both the project's strength and its weakness. The network gelled for the duration of the nutritional study with the 'Sisters of Nutrition', and was highly dependent on a uniquely Filipino mix of science, development and religion. Beyond the boundaries of the study (for which the convent, not the research laboratory, was central) the open, interdisciplinary environment soon broke down, and old disciplinary divisions and hierarchies quickly reasserted themselves. In the process, a familial environment tolerant of uncertainty was replaced by the strictures of formal institutional science, which soon severed connections between constituent disciplinary parts: discussions about 'bioavailability numbers', alternative milling strategies and multi-locational GxE trials now were conducted separately by different epistemic communities that rarely met.

Given the reassertion of the linear innovation model within the HarvestPlus Challenge Program (which, as discussed in Chapter 4, has since absorbed continuing iron rice research efforts at IRRI), this account highlights the shortcomings of such models, and associated notions of 'impact'. In this case, even a network with roots firmly planted in the Philippines was unable to pre-empt the complex of contextual factors that mediated processes of 'implementation' and ultimate impact. Reflecting on why the Philippines' first nutritional rice variety may have 'missed its moment'; one member of the assessment team suggested that the most significant lesson is that 'national priorities *matter*'.[74] A more nuanced reading of national politics of the time might have revealed the likely fate of problematic new policies, backed by a technocratic nutrition lobby, in the face of the high politics of rice self-sufficiency and national food security. In light of this, the vision built into HarvestPlus of directly targeting decontextualized 'populations at risk' ignores these critical dynamics at the national level and the ways in which these mediate and interact with local political dynamics in diverse socio-ecological settings.

3

An Institutional Model?
The Case of Golden Rice

Introduction

This chapter outlines a very different biofortification pathway, which began at around the same time as the iron rice project. The high-pro-vitamin A rice or 'Golden Rice' initiative has differed from early CGIAR biofortification research and the iron rice study in important ways. Unlike iron-enriched rice, Golden Rice, as its name suggests, is visibly different; the elevated levels of beta-carotene in Golden Rice lend it a distinctive yellow-orange colour. And while 'high-iron rice' had emerged from ongoing breeding programmes at IRRI and other public research institutions in Asia, Golden Rice was launched from a laboratory in Switzerland in 1999. Importantly, Golden Rice was transgenic[1] and, as such, became the focus of controversy and debate. The project had been funded (principally) by the Rockefeller Foundation under its International Program on Rice Biotechnology (IPRB) which had set out explicitly to link Northern 'advanced research institutions' with the expertise to conduct 'cutting edge' biotechnology research with institutions in 'rice-dependent' countries in the South.

The history of Golden Rice owes much to its early days as an international science policy controversy, and the ways in which its inventors and promoters framed its 'defence' and subsequent promotion. In particular, the early framing of Golden Rice as a scientific breakthrough whose realization was delayed, at significant humanitarian cost, by institutional factors alone has endured, effectively discouraging open discussion about uncertain aspects of this 'cutting edge' technology and its implementation. Such a polarized debate has focused the spotlight on the transgenic character of Golden Rice and associated intellectual property questions arising from its controversial acquisition by a private company – Zeneca (now Syngenta) – who subsequently granted it back to the inventors for free transfer to public research institutions in Asia through a novel institutional arrangement, the 'humanitarian licence'. This has been at the expense of a more comprehensive discussion about its merits as a nutritional intervention or – more fundamentally – as a staple food.

This chapter traces the Golden Rice pathway, from the international stage to its arrival at IRRI as the hub of the Golden Rice network – the 'technology holder' that disseminates Golden Rice materials to institutions in Asia for back-crossing into local varieties, overseen by a close-knit 'Humanitarian Board' based in Zurich. The transfer of materials occurred in the context of IRRI's uneasy relationship with its host country, the Philippines, where, since the 1980s, civil society groups had increasingly questioned the relevance of IRRI's research agenda. In light of this, 'defending' Golden Rice is a role that appears to have passed seamlessly from its inventors to IRRI, so that – despite its more obvious ownership of IR68144 (the high-iron rice variety that was the focus of the previous chapter) – biofortification is symbolized by the promise of Golden Rice in IRRI circles and beyond. This in turn has obscured the tangle of competing pathways that has characterized Golden Rice research in practice, enabling its presentation as a linear trajectory surmounting all obstacles.

Rice Biotechnology: Laying the Foundations

In 1984 the Rockefeller Foundation embarked on a 'highly speculative' venture to build an international biotechnology infrastructure and capacity around a single crop, rice (O'Toole et al, 2001, p1). The programme continued a tradition of investment by the Rockefeller Foundation in international agricultural science, beginning in China in the 1930s and the Mexican Agricultural Program (which evolved into CIMMYT) in the 1940s, and intensifying with the establishment of IRRI in 1960 and the formation in 1971 of the CGIAR donor group to support a network of agricultural research centres that became established during the Green Revolution (Normile, 1999; Lehmann, 2001).

In the early 1980s, foundation officials had turned their attention to the potential of agricultural biotechnology, concerned that private-sector investment would concentrate efforts towards the needs and markets of commercial farmers in the North. Assisted by Mahubub Hossain, an agricultural economist from IRRI, foundation staff conducted a two-year survey and analysis of the genetic prospects for the world's major food crops (Evenson et al, 1996; O'Toole et al, 2001). Over the following 17 years, the IPRB disbursed US$105 million,[2] an average of approximately US$6.2 million per year (O'Toole et al, 2001, p42), and in the process enrolled an international network of approximately 700 scientists in 30 countries, including 400 from Asian countries (Hindmarsh and Hindmarsh, 2002, p7).

The IPRB (1984–99) represented a point of departure from the foundation's earlier strategy of planting 'definitive centres' for crop research in developing countries where those crops were grown and consumed. Instead, the IPRB architects imagined an international map of rice biotechnology expertise in which the knowledge and skills necessary for 'cutting edge research' were located in Northern institutions, far from the projected needs of 'rice-dependent' populations. The first phase of the programme focused on building the 'scientific basis for "rice biotechnology" as we know it today' (O'Toole et al, 2001, p1) with its roots firmly in 'advanced research institutions' in the North:

> Early successes were the first DNA molecular marker map of rice, the regeneration and transformation of rice, the use of rice pest genomic information to unravel age-old riddles of host-plant resistance, and numerous other discoveries that changed the way rice geneticists viewed breeding objectives such as insect resistance, abiotic stress tolerance, and hybrid rice. These discoveries culminated in the revelation of rice's pivotal genomic position in the evolution of cereal species. (O'Toole et al, 2001, p1)

In its second phase, the programme shifted to an emphasis on technology transfer, specifically the 'transfer of resulting technologies to institutions in rice producing and consuming countries' (O'Toole et al, 2001, p1) or 'rice-dependent' countries (Herdt, 1996, p17) in the South. This was achieved through an 'international collaborative research-cum-training' model (O'Toole et al, 2001, p1):

> The Foundation's program management sought to support further technology generation and application while promoting the program's greatest asset, international collaborative research-cum-training. This 'win–win' component of the program linking fledgling national rice biotechnology efforts directly to advanced research institutes in the United States, Europe, Japan, and Australia became the hallmark of the Foundation's management strategy ... The successful linkage of research in cutting-edge biotechnology with the training of rice scientists often produced long-term collaborative relationships that outgrew dependence on Foundation support and continue today (such as the IRRI-managed Asian Rice Biotechnology Network). (O'Toole et al, 2001, p1)

Through this programme a new generation of Asian scientists were socialized into an evolving international community of scientists through doctoral and post-doctoral scholarships and membership of international research teams. The success of this model in establishing a critical mass of talent at institutes such as PhilRice in the Philippines is considered a case in point:

> Of eight senior researchers at the Philippine Rice Research Institute at Maligaya, for example, five earned their doctorates with Rockefeller grants and one received a postdoctoral career development grant for collaboration research with an advanced lab. 'Without the

support of The Rockefeller Foundation, it would have been almost impossible for us to build this capability', says Leocardo Sebastien, the institute's deputy director for research.[3] (Normile, 1999, p1468)

One project funded through the IPRB during the 1990s was a Swiss-German initiative to create high-pro-vitamin A rice, or 'Golden Rice' as it is now popularly known. The achievements of the Zurich-based research team exceeded the expectations of all concerned so that, despite the original imperative behind the IPRB to raise yields (Herdt, 1996), it was this result that foundation officials singled out as 'the program's greatest achievement':

> Robert Herdt, Rockefeller's Director for Agricultural Sciences, ticks off an impressive list of successes, including rice lines that are tolerant of high-aluminium soils and the identification and transfer of a gene associated with resistance to bacterial blight. The program's greatest achievement, however, has been support for work that incorporated the synthesis pathway for beta-carotene, the precursor to vitamin A, into rice. Herdt says that an estimated 400 million people dependent on rice suffer vitamin A deficiency, with its associated vision impairment and disease susceptibility. (Normile, 1999, pp1468–9)

In order to understand how this came about, it is necessary to trace developments in the field of nutrition that have led vitamin A deficiency to have such a high profile as a global public health priority.

Vitamin A Deficiency: Construction of a Public Health Problem

As discussed in Chapter 1, during the 1980s findings emerged which shifted the focus of international nutrition to micronutrients. While there was already an understanding of the importance of these nutrients, the new studies highlighted findings that demonstrated clearer *economic* consequences of micronutrient deficiencies (Horton and Ross, 2003; Gillespie et al, 2004). In particular, Sommer's findings released in 1984 (Sommer et al, 1984), the same year that the Rockefeller Foundation launched the IPRB, showed that 'provision of vitamin A had benefits beyond prevention of known deficiency diseases; it also affected mortality from other causes. This research altered the cost–benefit calculation for vitamin A intervention' (Gillespie et al, 2004, p82).[4]

Similarly, while the phenomenon of iodine deficiency had been recognized for some time – UNICEF had promoted salt iodization since the 1960s – 'the breakthrough development was the new concept of iodine deficiency disorders (IDD), where even mild degrees of iodine deficiency caused functional deficits, especially in intelligence' (Gillespie et al, 2004, p95). Furthermore, in the case of iron deficiency, the effects of anaemia on 'productivity and cognitive function' (Gillespie et al, 2004, p96) were well known since the 1970s. Together, these three micronutrients[5] became the focus of a series of international initiatives and conferences, in particular the 'Ending Hidden Hunger' conference held in Ottawa, Canada, in 1991 (dela Cuadra, 2000). Micronutrient interventions were increasingly viewed as both cost-effective preventative healthcare (Darnton-Hill, 1998) and investments in future productive and innovative capacity (Slingerland et al, 2003). As discussed in Chapter 1, these arguments gained further currency within the MDG framework, attracting endorsement from an influential group of economists through the 'Copenhagen Consensus' initiative (Behrman et al, 2004).

Of these three micronutrients, however, it is vitamin A that has retained the highest profile. With the association of its deficiency with blindness, particularly in children, vitamin A deficiency has remained the most obvious and emotive of the micronutrient deficiencies. The primary intervention has been equally high profile – supplementation programmes piggybacked on to national immunization days (NIDs). While noted for their high coverage, concerns remains about the sustainability of these programmes, particularly in the 'post-NIDs era' (IVACG, 2003, p7). In contrast, the case of iodine deficiency and salt iodization has been less contentious and iron deficiency – while arguably the most pressing micronutrient problem on a global scale (Horton and Ross, 2003) – suffers both from the invisibility of its lasting effects and its exemplification of the complexity of interacting socio-economic and biological factors underpinning single indicators of nutritional status:

> The most obvious deficiencies are the easiest to tackle, and those more difficult to observe tend also to be the most difficult to prevent ... Thus, vitamin A deficiency is known to cause blindness and its prevention demonstrably protects sight, as well as substantially reducing death in children (and probably mothers) in areas of deficiency. Moreover, prevention can be readily achieved through an infrequent supplement, whose distribution can be monitored in a straightforward manner. Similarly, iodine deficiency has visibly distressing results, such as cretinism, which is easily prevented, at least in principle, by iodized salt, and the use of iodized salt can also be simply tracked. Conversely, iron deficiency anaemia is less readily observed and less easily prevented. (Mason et al, 2001, p2)

Nevertheless, global efforts to reduce vitamin A deficiency (VAD) have been characterized in terms of broad improvements through established programmes benefiting from economies of scale (for example, see Mason et al, 2001, on cost savings from UNICEF's consolidated purchasing of vitamin A capsules). The authors of *The Micronutrient Report* present an optimistic, if tentative, assessment of the global impact of ten years of VAD programming:

Trends in the prevalence of clinical vitamin A deficiency (mainly Bitot's spots) were assessed by comparing repeat national survey results, when these were available ... Except in Niger and Nepal, a pattern of improvement is evident ... For subclinical vitamin A deficiency assessed by serum retinol levels <0.7 μmol/L the picture is less clear ... A judicious conclusion may be that the subclinical vitamin A deficiency results, which are almost always taken from surveys conducted independently of clinical assessments, support the idea that a significant and broad improvement is under way. (Mason et al, 2001, pp21–3)

The Philippines National Vitamin A Supplementation Program (NVASP) is considered 'one of the oldest, most mature and comprehensive of its kind' (Fiedler et al, 2000, p223). VAD was identified as a priority area in the Philippines during the 1970s and addressed through a monosodium glutamate (MSG) fortification programme (Solon et al, 1979). Helen Keller International (HKI), an international NGO specializing in VAD programming, arrived in 1975 and has been 'a major presence' since 1986, playing 'a critical role in the implementation of the first vitamin A capsule distribution campaign'.[6] In 1993 a nationwide programme was launched with strong political support from the Ramos administration, in the wake of the 'Ending Hidden Hunger' conference (dela Cuadra, 2000). This has been complemented by a range of public-private initiatives fortifying staples such as cooking oil, margarine, sugar and wheat flour with vitamin A (Solon, 2000).

However, one leading nutritionist has drawn attention to a sequence of events which led to the construction of VAD as a public health problem of national importance, highlighting how this was contingent on the nature and timing of policy shifts and interventions of international agencies, so that a reading of data then available may, at a different time, have led to a different set of conclusions and outcomes:

At the time the plan was formulated, the only data that was available for consideration were the 1986 nutrition surveys, 1987 studies in depressed areas in one municipality and one province and 1987 national nutrition survey. Data ... indicate[d] VAD to be a public health problem in two out of six regions and in the two small, depressed areas. The national nutrition survey failed to establish VAD as a problem of public health significance in the country based on both biochemical and clinical indicators. The food consumption data supported these findings. In the 1987 national survey, the mean one day vitamin A intake of 6 month– 6 year old children was 89.4% of their RDA. It is well recognized that a nutrient deficiency as a public health concern on the local level should not be ruled out based on aggregate national results. But should the existence of a significant public health problem deserving of nationwide intervention be established on the basis of a few localized studies in depressed areas, particularly when national surveys do not indicate such a deficiency based on biochemical and clinical indicators with supportive evidence from dietary intake? (Florencio, 2000, p4)

Accounts such as this are illustrative of tensions that exist between representations of global coverage and trends and the contingencies and ambiguities that characterize national and local contexts, even in countries with 'mature and comprehensive' nutrition programmes. Furthermore, these policy uncertainties coexist with a still evolving body of knowledge about micronutrients and human nutrition. For example the International Vitamin A Consultative Group (IVACG)[7] revised the recommended vitamin A supplementation dose in 2001 (Sommer and Davidson, 2002); and the bioavailability of vitamin A from carotenoids in plant foods has been the subject of controversy and debate in recent years (Brown et al, 2004; IVACG, 2003; Van Lieshout et al, 2002; West et al, 2002).

Nevertheless, networks formed around VAD as a global problem and pharmaceutical supplement distribution as the solution extended to countries such as the Philippines. Scattered findings were aggregated and black-boxed as evidence of a national public health problem, despite inconsistencies with established national survey data. At the same time, these processes of black-boxing at the national level served to consolidate the construction of a 'global' problem worthy of increased levels of support. It was into this ambiguous terrain that in 1999 Golden Rice was presented as a breakthrough in the global battle with VAD. The next section traces the early history of the research, from its exploratory stages as a project for which those involved had until then had relatively low expectations.

Golden Rice: A Scientific Breakthrough

In the early 1990s one of many funding applications submitted to the IPRB was a joint application from Ingo Potrykus of the Institute of Plant Sciences at the Swiss Federal Institute of Technology (Eidgenössische Technische Hochschule – ETH) in Zurich together with Peter Beyer at the University of Freiberg Centre for Applied Biosciences, Germany. At the time, Potrykus, trained in plant genetics, was nearing the end of a scientific research career in the prestigious Zurich institute.[8] Though a laboratory scientist for much of his career, he has described himself as, first and foremost, a biologist more interested in science that yields 'practical benefits' than 'science for the sake of pure knowledge': 'Though his reputation is based on it, Potrykus has never seen himself as a genetic engineer. He says he is more of a *"Wald and Wiesen"* biologist, a fields-and-meadows type of guy who prefers studying birds, insects and plants over any indoor lab environment.'[9]

Potrykus and Beyer's project aimed 'to genetically engineer the pro-vitamin A pathway into the rice endosperm' (Potrykus, 2001, p1157). At this stage the recently established (1985) Plant Sciences Institute was already in receipt of a number of research grants from the Rockefeller Foundation. Over the next few years, however, a set of connections were established which would provide a core of support for a more ambitious project. Along with Potrykus and Beyer, this emerging network included Swapan Datta, a new recruit to Potrykus's research team. Datta introduced Potrykus to Gary Toenniessen of the Rockefeller Foundation, who 'responded with the organization of a brainstorming session in New York'.[10] According to Potrykus, 'The verdict of this initial session was that such a project had a low probability of success, but that it was worth trying because of its high potential benefit' (Potrykus, 2001, p1158).

This was the beginning of a long association between Toenniessen and the two Golden Rice inventors that would outlive the IPRB, leading one colleague to dub Toenniessen 'the father of Golden Rice'.[11] Trained as a microbiologist, Toenniessen's first encounter with the Rockefeller Foundation was as a postdoctoral fellow. He later joined the foundation as a programme officer and was responsible for developing and implementing the IPRB from 1985 onwards.[12] Overseeing the programme was its main architect, Robert Herdt, an agricultural economist who had already served at senior levels within the CGIAR and its founders, the Rockefeller and Ford foundations, for many years. Starting with the Ford Foundation's integrated rural development programme in India in the 1960s, Herdt then went to IRRI for ten years (1973–83) as head of the Economics Department, followed by three years as a science advisor to the CGIAR secretariat, evaluating centre programmes. Thereafter he spent 17 years with the Rockefeller Foundation, as director of Agricultural Sciences (for 12 years) and then vice president for Programme Administration.[13] This continuity in leadership of the foundation's agricultural programming in general, and the IPRB in particular, was thus extended by Herdt's long association with the CGIAR and both of its parent foundations. When Herdt left the Rockefeller Foundation in 2003, Toenniessen remained and continued to champion the project.

In 1993, with '$100,000 in seed money from the Rockefeller Foundation, Potrykus and Beyer launched what turned into a seven year, $2.6 million project, also backed by the Swiss Government and the European Union'.[14] In 1999, the first breakthrough was achieved: 'Xudong Ye of my laboratory did the crucial experiment: cotransformation with two *Agrobacterium* strains containing all the necessary genes plus a selectable marker. The resulting yellow-colored endosperm contained provitamin A and other terpenoids of nutritional importance and to everybody's surprise demonstrated that it was possible to engineer the entire biochemical pathway' (Potrykus, 2001, p1158; see also Ye et al, 2000).

These 'surprising' results were emerging just as the Rockefeller Foundation was closing down the IPRB. At this stage, what later came to be called the 'Golden Rice prototype' (Al-Babili and Beyer, 2005, p565) contained 1.6 µg/g provitamin A in the endosperm (Potrykus, 2001, p1158). In 1999 Potrykus, at

65 about to retire from ETH, and Beyer announced their achievement (Nash, 2000, pp38–46).

A Science Policy Controversy

The status of 'Golden Rice' as an international science policy controversy is well documented (Nash, 2000; BIOTHAI (Thailand) et al, 2001; Potrykus, 2001; Jasanoff, 2005). In 2000 *Time* magazine featured Golden Rice on its cover with the announcement: 'This rice could save a million kids a year' and an article presenting aspects of a debate that was becoming increasingly polarized, as not only the efficacy and appropriateness of the new technology, but also the intent of its promoters was called into question (Pollan, 2001). In the process, promoters and defenders of Golden Rice each began constructing their own 'Golden Rice Tale'.[15]

Robert Derham of Checkbiotech has suggested that 'If Golden Rice could speak, it would probably tell its story' through Potrykus, Beyer and Adrian Dubock of Syngenta.[16] The process by which Golden Rice was transformed from a public-sector research project to a public-private partnership in which Syngenta (then Zeneca)[17] was to take a pivotal role, led critics to assert that biotech companies were employing Golden Rice as a 'Trojan horse' to gain public acceptance for genetically modified (GM) crops.[18] Even the original sponsors have acknowledged, with some irony, that a product developed through the IPRB 'with no financial backing from the private sector ... has not stopped companies from using it as a "poster child" in their promotional efforts' (Herdt et al, 2005, p5). In the process, discussions about the merits and shortcomings of Golden Rice were absorbed into broader debates about GM crops, in which malnourished children in developing countries were cast as unwitting victims of a transatlantic dispute driven by elite concerns (Pollan, 2001; Potrykus, 2001).

While a product of crop genetic engineering, the benefits claimed for Golden Rice were nonetheless *nutritional*: consuming Golden Rice would improve nutrition and therefore health status. How did nutritionists respond to such claims? Dr Marion Nestle, a professor in the Department of Nutrition and Food Studies at New York University, was unambiguous:

> Consideration of basic principles of nutrition suggests that rice containing beta-carotene is unlikely to alleviate vitamin A deficiency. To begin with, the bioavailability of beta-carotene is quite low – 10% or less by some estimates. To be active, beta-carotene – a pro-vitamin – must be split by an enzyme in the intestinal mucosa or liver into two molecules of vitamin A. Like vitamin A, the pro-vitamin is fat-soluble and requires dietary fat for absorption. Thus, digestion, absorption and transport of beta-carotene require a functional digestive tract, adequate protein and fat stores, and adequate energy, protein and fat in the diet.

Many children exhibiting symptoms of vitamin A deficiency, however, suffer from generalized protein-energy malnutrition and intestinal infections that interfere with the absorption of beta-carotene or its conversion to vitamin A. In numerous countries where vitamin A deficiency is endemic, food sources of beta-carotene are plentiful but are believed inappropriate for young children, are not cooked sufficiently to be digestible, or are not accompanied by enough dietary fat to permit absorption. In addition to doubts about cost and acceptability, biological cultural and dietary factors act as barriers to the use of beta-carotene, which explains why injections or supplements of pre-formed vitamin A are preferred as interventions. The extent to which the beta-carotene in golden rice can compensate for these barriers is limited. (Nestle, 2001, pp289–90)

Critics aligned with the 'anti-GM' lobby, however, focused instead on the modest (absolute) levels of beta-carotene contained in Golden Rice. To make their point they employed the memorable caricature of a child attempting to consume the several kilograms of Golden Rice that would be necessary to satisfy their daily vitamin A requirement (Pollan, 2001) (see Figure 3.1). Increasingly on the defensive, Potrykus drew attention to what he clearly saw as a catch-22: nutritional assessment was indeed necessary, but could not be carried out 'with the few grams of rice' in the greenhouse (Potrykus, 2001, p1158). At this time, however, Switzerland (and much of Europe) had placed a *de facto* moratorium on the commercialization of GM crops,[19] so field planting was unlikely.

The rapid polarization and escalation of debates at this time has had a lasting effect on the framing of Golden Rice as a research project and as a potential solution to VAD as a global problem. In particular, an initial posture of defence, in reaction to an unexpected level of opposition, has been instrumental in shaping the subsequent Golden Rice pathway. Ingo Potrykus, who since his retirement has taken on a role as the 'voice' of Golden Rice, defines its pathway in terms of a scientific breakthrough whose realization has been impeded by a series of 'roadblocks',[20] stalling its otherwise inevitable dissemination and acceptance. A linear process was envisaged, in which successive hurdles – technology transfer, regulatory approval and consumer acceptance – were to be negotiated one by one (Potrykus, 2001).

Freedom to operate

The first 'roadblock' encountered by the inventors arose when they discovered that, in the process of their research, they had infringed private intellectual and technical property rights. At this point, the problem was defined in terms of 'freedom to operate' – how to transfer the technology to institutions such as

Figure 3.1 *Eat your dinner: Activist poses as a child faced with the volume of Golden Rice necessary to meet daily vitamin A requirement*

IRRI and NARS in such as way that they could continue with adaptive research and field testing. As Potrykus recalls:

> Peter Beyer had written up a patent application, and Peter and I were determined to make the technology freely available. Because only public funding was involved, this was not considered too difficult. The Rockefeller Foundation had the same concept and the Swiss Federal Institute of Technology (Zurich) supported it, but the European Commission had a clause in its financial support to Peter Beyer, stating that industrial partners of the 'Carotene Plus' project, of which our rice project was a small part, would have rights to project results. (Potrykus, 2001, p1158)

In 2000 the Rockefeller Foundation commissioned the International Service for the Acquisition of Agri-biotech Applications (ISAAA) to conduct a 'freedom to operate review' of Golden Rice. This review identified a total of 70 patents and 15 technical property (TP) agreements – usually referred to as material transfer agreements or MTAs – applicable to Golden Rice. Based on this assessment, six options were outlined, as follows:

1   invent around current patents;
2   redesign constructs and 'wherever possible synthesize own genes to reduce reliance of TP of others';
3   persuade intellectual property (IP)/TP owners to relinquish claims (a 'humanitarian use' option);
4   ignore IP/TP rights;
5   seek licences for all IP/TP; and finally
6   a recommended option combining 2, 3 and 5 (Kryder et al, 2000, ppvii–viii).

Notably, given that subsequent debates have highlighted the profusion of patents, the authors stated that the second option, to invent around MTAs, 'would almost certainly be the approach favoured by any company as the TP issues are potentially the most difficult ones to resolve' (Kryder et al, 2000, pvii). While MTAs are non-territorial, the enforceability of the patents always depends on national legal frameworks (Kryder et al, 2000, ppvi–vii). In the case of Golden Rice, the NGO, GRAIN, has observed that: 'Of the 60 countries with vitamin A deficiency – which Golden Rice is supposed to address – only 25 could possibly honour any of the patents involved. And in those countries, only 11 of the patents could constrain the projects locally.'[21]

In retrospect, ISAAA recognizes 'that the initial analysis was an exaggeration of the amount of IP [intellectual property] issues in the product' noting that 'this initial analysis raised concern that it would make it almost impossible for products like this to move, [and] unduly raised unnecessary alarms'.[22] Nevertheless, critical decisions were made at that time on the basis of this initial assessment which would set the course of project in a particular direction, as the following discussion illustrates. In particular, the reference in the ISAAA review to a 'humanitarian use' option referred to discussions that were already taking place between Golden Rice inventors and Adrian Dubock, Zeneca's commercial biotechnology manager, as one of the main patent holders. The concept of humanitarian use, which could be granted through a legal instrument called a 'humanitarian licence' would enable Zeneca and other companies to relinquish their IP claims on Golden Rice in instances where it was for humanitarian purposes, yet retain those rights in other circumstances where it might be profitable. 'Humanitarian use' in this case would be defined by an annual income ceiling for farmers receiving the technology. Crucially for IRRI (and its network of NARS in Southeast Asia) this ceiling could be set at a level which would make the technology freely available to all farmers in those target countries (Potrykus, 2001).

These negotiations accelerated the selection of the third option presented in the ISAAA review. The ownership of Golden Rice was transferred from the public sector 'via a small licensing company (Greenovation, Freiberg, Germany)' to Zeneca/Syngenta as the primary (but not exclusive) patent holders (Potrykus, 2001). As Toenniessen, a long-term champion of the project, explained:

> Drs Potrykus and Beyer have the rights under this agreement to share Golden Rice with public-sector rice breeding programs to generate new Golden Rice varieties for use by resource-poor farmers in developing countries, defined as farmers generating less than US$10,000/yr. income from Golden Rice. This is known as the Humanitarian Project. Zeneca has retained all commercial rights in all countries and will donate support to the inventors in the Humanitarian Project. (Toenniessen, 2000, p5)

Under this agreement Zeneca took on the role of negotiating the IP/TP maze on behalf of the Golden Rice inventors and in the process other companies, including Monsanto, also relinquished IP claims:

> It turned out that our agreement with Zeneca and the involvement of our partner in Zeneca, Adrian Dubock, were real assets in developing the humanitarian aspect of the project. Adrian was very helpful in reducing the frightening number of IPRs and TPRs. He also organized most of the free licenses for the relevant IPRs and TPRs such that we are now in the position of granting 'freedom to operate' to those public research institutions in developing countries to proceed in introducing the trait into local varieties. (Potrykus, 2001, p1159)

Potrykus's account maintains the image of a linear pathway in which a series of hurdles are overcome and makes no reference to the relative merits of the various options presented in the ISAAA review, which might have led to alternative pathways. 'Having overcome the scientific problems, and having achieved freedom to operate, leaves technology transfer as the next hurdle,' announced Potrykus (2001, p1159). What appears to have been evolving, at that stage, was the envisioning of a linear innovation/diffusion pathway consistent with Rogers's linear model for innovation (Rogers, 2003), but with an important difference in that each successive milestone along the linear path presented some form of potential difficulty – in terms of regulatory complexity or purposeful resistance – so that new types of governance arrangements were deemed necessary.

In this case, the transfer process would be overseen by a 'Humanitarian Board'; initially a core group chaired by Potrykus and including Beyer and Dubock:

> Although we have had requests from many institutions in many countries, we believed it would be unwise to start the technology transfer on too large a scale. To aid in this endeavor, we have established a 'Golden Rice Humanitarian Board' to help make the right decisions and to provide secretarial support. Again, our decision to work with Zeneca was extremely helpful. Adrian Dubock was willing to care for the task of the secretary. We have additional invaluable help from Katharina Jenny from the Indo-Swiss Collaboration in Biotechnology. (Potrykus, 2001, p1159)

The core group driving the project at this time were Potrykus, Beyer, Dubock and Toenniessen. The caption for a picture of the four men, posted on the Golden Rice website,[23] describes them as the 'fathers' of Golden Rice. Also on the website are individual biographies of all the Humanitarian Board members. The biographies for Potrykus, Toenniessen and Dubock refer to these core members as, respectively, the 'engine', the 'visionary' who started it all, and the 'architect of the public-private partnership that has made Golden Rice possible'. The humanitarian credentials of Dubock, a zoologist and former rodent specialist turned 'career pharmaceutical executive',[24] are given particular emphasis: 'Through his work Adrian has been to many countries, and apart from the beauty, he saw many things that were not quite right. This motivated him to do whatever he could from within his position, converting him into a Syngenta ambassador to the world.'

Over time, the board has extended to include a broad range of disciplines, donors and other supporters, and now has 13 members. At the time of writing, these included representatives from universities in Europe and the US (in Zurich, Freiberg, Hohenheim, Cornell and Boston); international agencies in Washington DC (the World Bank, USAID, and HarvestPlus – represented by Howarth Bouis); the Indian Department of Biotechnology; and IRRI in the Philippines. A photograph of the Humanitarian Board, posted on the Golden Rice website at the time of the first Golden Rice field trial, in Louisiana in September 2004, reveals a group of nine men and one woman, overwhelmingly white and middle-aged, lined up in front of a row of Wellington boots, presumably about to venture into the field to view the harvest.[25] One wonders how their individual and collective understandings of their 'humanitarian' mission would compare with that of the iron rice 'family', which, as described in the previous chapter, was so firmly anchored in a shared socio-cultural world.

Proof of concept

Debates at this time focused attention on the institutional arrangements necessary to facilitate technology transfer and 'freedom to operate', while maintaining an implicit assumption that the technology, once in the hands of the plant breeding institutions in Asia, would prove appropriate and adaptable to local needs and conditions. While questions such as those raised by Nestle remained largely unanswered this was not considered problematic due to a subtle shift in the framing of what had so far been achieved. The Golden Rice project was reframed as 'a *proof of concept* study', which had simply 'shown that [it was] *possible* to establish a biosynthetic pathway *de novo* in rice endosperm, enabling the accumulation of pro-vitamin A' (Beyer et al, 2002, p510s, emphasis added).

The next step was therefore to transfer the technology back to the public sector, to institutions able to introduce the trait into local varieties. Meanwhile, the role of Zeneca, the new owner of Golden Rice, had broadened from one of facilitating access to intellectual property to providing guidance on future scientific and product development:

> Both the inventors and the Foundation wanted thorough testing for biosafety and nutrition, but such testing is expensive. They eventually concluded that the best and quickest way to overcome the IP constraints and test the product was to enter into a partnership with a company, Zeneca, that already had strategic and research interest in both rice and nutritional enhancement of food and consequently access to a large IP portfolio relevant to modifying the carotenoid biosynthetic pathway in plants, plus extensive experience in biosafety testing of crops and foods ... Under the partnership agreement, Zeneca will help the inventors further modify their research product to produce a commercially viable variety of Golden Rice for dissemination to breeding programs, and will also facilitate biosafety testing and nutrition studies. (Toenniessen, 2000, p4).

'Proof of concept' is a term familiar to those working in business development and marketing, referring to the initial feasibility testing of a business idea. Golden Rice appears to be the first example of its use within an agricultural research context – which seems surprising now the term has become ubiquitous within biofortification circles and has found its way into the CGIAR lexicon (for example, see Science Council and CGIAR Secretariat, 2004, p7).

As a boundary term (Gieryn, 1999), the reframing of Golden Rice research, thus far, as a proof of concept study drew a line between the laboratory-based endeavours of the scientists and the various arguments as to the efficacy, effectiveness and appropriateness of their invention, even though these debates had, in the first place, originated in the scientists' own claims for the significance of their findings. This provided, in the midst of heated debates, an 'escape hatch' (Clay and Schaffer, 1984b, p192) through which the Golden Rice inventors could distance themselves from the claims and counter-claims now proliferating, and from their possible consequences, and at the same time pursue their argument that Golden Rice research should continue.

In the context of a science policy controversy in which various groups were contending that Golden Rice would not 'work', the response from the investors that they were only claiming 'proof of concept' appears, in retrospect, as a master stroke: it lowered the bar for success just far enough for the 'breakthrough' claim to be sustained, while acknowledging that the work was incomplete and therefore in need of further resources. Obscured within an apparently neutral technical statement was an implicit argument that further research and testing of Golden Rice should be conducted and supported, a claim that required no further justification. Concerns such as those of Nestle (2001) would be dealt with in due course, once the regulatory environment was more conducive. Crucially, the use of the term effectively framed out questions that might have challenged the kind of assumptions that had enabled promoters to present Golden Rice as a product whose delay was causing widespread suffering (Nash, 2000).

Granting Access, Keeping Control

On 22 January 2001, Golden Rice materials were delivered to IRRI. A joint press release from IRRI, the Rockefeller Foundation and Syngenta introduced the Humanitarian Board as 'an exciting new type of public-private sector collaboration, formed specifically to further an important scientific breakthrough in the development field'. By this time the Humanitarian Board had now been extended from the original core group to include representatives from IRRI, the World Bank and Cornell University. Its mandate had now been extended to cover 'four principle aims':

- to support the inventors in making 'Golden Rice' freely available to those that need it, consistent with the highest standards of safety assessment;
- to ensure the proper investigation of 'Golden Rice' as one potential solution to Vitamin A Deficiency;
- to support individual developing countries and their national research institutes as they assess their interest in 'Golden Rice'; and
- to facilitate information-sharing between 'Golden Rice' projects in different parts of the world.[26]

At the same time, plans were under way to transfer the materials to other institutions in Asia. Initially this included India, Vietnam, China and Indonesia:

> In addition, facilitated by the Indo-Swiss Collaboration on Biotechnology, further research and development of Golden Rice in India is being pursued in collaboration with national research institutes. Dr Hoa, a Vietnamese visiting scientist, has transformed several local varieties. She will take these seeds back to Vietnam to conduct further research there, in accordance with a sublicense agreement with the Cuu Long Delta Rice Research Institute. A possible transfer to China is currently being discussed with the Chinese Ministry of Science and Technology; the

Minister of Agriculture of Indonesia also has expressed interest in entering into similar discussions. (Beyer et al, 2002, p509s)

The question for IRRI at this time was how to reconcile this new public-private partnership arrangement with its mandate to generate international public goods:

> We had to make them understand that we are an institute committed to produce public goods, and therefore anything we receive, even in collaboration with other parties, will cascade into our national partners. Nothing ever stays here in Los Baños, because we would not be true to our commitment to produce international public goods. In other words, if they execute an agreement with us, we want our cards to be laid on the table. We are an institute committed to produce international public goods, especially in this part of the world, in Asia. We cannot allow restrictive agreements, and exclusive agreements, so that principle we held onto, until the concept of a humanitarian license came in.[27]

This principle applied not only to the free *distribution* of technologies and products, but also to the freedom of IRRI and NARS to undertake adaptive research. In this case the definitions of 'humanitarian use' and 'resource poor farmer' contained in the licence were deemed 'more than adequate' to ensure IRRI did not divert from its mandate:

> One of restrictions we thought we would be very careful with [was] the extent of distribution: because it's humanitarian we should be able to distribute it as widely as possible. And we thought the definition of a resource-poor farmer in that license was more than adequate to cover the kind of clients we would like to share the technology with ... We don't want to just be confined to use the [material] *per se* and just conveying it *per se*, because we know that, in the first place the material was in a general background that was not going to be useful for most of Asia – *japonica* background – so they agreed. And I think it's just logical that when you get a humanitarian license it's not just a matter of allowing its distribution to as wide a group as possible, but also allowing for adaptive research to be undertaken because, well, agriculture by its very nature depends on the interaction of the gene and the environment.[28]

These accounts indicate that IRRI officials went to some lengths to resolve potential conflicts between the new intellectual property regime within which the humanitarian licence was developed and IRRI's *raison d'être* as a public goods research institute used to disseminating its research output freely throughout its network of NARS in the region. However, though these initial, formal agreements appear to have been satisfactory, the question remained as to how these new arrangements might play out in practice once adaptive research was under way.

Golden Rice research has subsequently followed two parallel but less than equal pathways. Scientists in Asia – notably Swapan Datta, by then leading rice biotechnology research at IRRI, and Hoa Tran Thi Cuc at the Cuu Long Rice Delta Research Institute in Vietnam – began using the prototype Golden Rice materials transferred to IRRI in January 2001 as their starting point. Their achievements included transformations of the Golden Rice trait into *indica* varieties, including IR64 and IR68144,[29] which were successful, but they did not increase the level of pro-vitamin A (Datta et al, 2003; Hoa et al, 2003). In the process they achieved a reduction in the number of recombinant events and the replacement of the antibiotic selection system in order to create a product 'more amenable to deregulation' (Datta et al, 2003; Hoa et al, 2003) and completed preliminary retention studies (Datta et al, 2003).

These introgression activities were coordinated through a 'Golden Rice network', with IRRI as its 'hub'.[30] IRRI recruited Gerard Barry,[31] formerly director of Research, Production and Technical Cooperation with Monsanto and, for many years, a key figure in Monsanto's rice programme,[32] to coordinate the Golden Rice network (and represent the network on the Humanitarian Board in an ex-officio capacity). By this time, the network had expanded to include institutes in the Philippines (PhilRice in addition to IRRI), Vietnam (Cuu Long), India, Bangladesh, China, Indonesia and Germany (University of Freiberg).[33] While working with some of its traditional NARS partners, however, the *modus operandi* of the Golden Rice network suggests a more constrained flow of information and materials than had previously been the case:

> Whatever materials are available at IRRI are accessed for adaptive research by the different countries, and there is still no cross country relationship, no cross country activity, it's a hub, and each spoke independently relating to IRRI. VietNam will relate to IRRI, India will relate to IRRI, Philippines will relate to IRRI, Bangladesh will relate to IRRI. But there is no relationship between Vietnam and India, Philippines and India.[34]

At the same time, a second research pathway was under way in the laboratories of Syngenta. Continuing with *Japonica* varieties, scientists at Syngenta were successful in raising levels of beta-carotene, first to 6 μg/g in events known as 'Golden Rice 1' or 'GR1' (Al-Babili and Beyer, 2005, p568), then to 37 μg/g in events known as 'Golden Rice 2' or 'GR2' (Paine et al, 2005, p1).[35] GR1 events resulted from the replacement of the CaMV 35s (cauliflower mosaic virus) promoter with an endosperm-specific promoter, producing plants both higher in beta-carotene and 'ultimately devoid of a selectable marker gene' so more 'suitable for deregulation' (Al-Babili and Beyer, 2005, p568). Submitted for field trials in the US in 2004, the GR1 events in a

Cocodrie background represent 'the best investigated events available to date' (Al-Babili and Beyer, 2005, p568), with bioavailability studies then planned for 2006 (Al-Babili and Beyer, 2005, p570).[36] The dramatic increases in beta-carotene levels in the later GR2 events in a *Kaybonnet* background were the result of substituting a maize gene (encoding phytoene synthase or *psy*) for the daffodil gene used previously (Paine et al, 2005).

The controlled transfer of GR1 and GR2 to the Golden Rice network, via the Humanitarian Board, then IRRI, is indicative of the hierarchical arrangements that were emerging, in which IRRI increasingly played the role of 'gatekeeper'[37] on behalf of the Humanitarian Board, as the following paragraphs illustrate. In 1994, Syngenta issued a press release to announce the transfer of GR1 to the Humanitarian Board: 'Syngenta announced today the donation of new Golden Rice seeds and lines to the Golden Rice Humanitarian Board. The donation follows the successful completion of the first Golden Rice field trials and harvest in the USA last month; it also marks World Food Day on 16 October and the UN's International Year of Rice this year.'[38]

By this time, scientists at IRRI and Cuu Long had been working with the Golden Rice prototype for a number of years. Nevertheless, there was a 'consensus' to shift research efforts towards the Syngenta GR1 events, as one network member recalls: 'Potrykus instructed our director to destroy the materials ... So it's totally gone ... I destroyed all the materials ... They started with this new one ... They use *Cocodrie* and IR64 ... they say [it is] more adaptable to Asian countries'.[39]

As with the prototype, however, GR1 results were still only partially understood; and associated uncertainties were again carried downstream and incorporated in the adaptive breeding stage:

> These GR1 events showed significantly higher carotenoid con-
> tents, up to 6.0 µg/g, whereas the public sector events showed a
> maximum of 1.6 µg/g. Explanations for this difference might lie in
> the use of a different promoter, the different biochemical back-
> ground of the cultivar used, or positional effects and, thus, in the
> larger number of independent transformation events produced.
> *Subsequent introgression into different cultivars of rice, which is
> currently underway, is expected to shed light on some of these ques-
> tions.* (Al-Babili and Beyer, 2005, p568, emphasis added)

In March 2005 the Humanitarian Board announced the GR2 results to the press.[40] By mid-2007, it had been transferred to IRRI, but had yet to be made available to other GR network members, who continued their introgression activities using GR1. As one IRRI scientist remarked, 'we are ahead of them, so can give guidance. And we can also give them our finished product.'[41] Meanwhile, while absolute beta-carotene levels in the GR2 events appeared markedly higher, the reliability of these results remained uncertain:

Although GR events have been produced in several *indica*, *javanica* and *japonica* cultivars, the influence of the different biochemical backgrounds of these different genotypes has not yet been studied. The high carotenoid content observed with *Cocodrie* and *Kaybonnet*, for instance, could also be the result of the much larger number of events created with these two cultivars, which would adequately account for positional effects impacting on gene expression. Introgression using marker-assisted selection is currently being carried out in Asia, which should provide information to make this distinction. In any case, cross-breeding is required to introgress the GR trait into cultivars of rice, mainly *indica*, that are locally adapted in areas with VAD. (Al-Babili and Beyer, 2005, p571)

These accounts indicate research processes that were continually opening up new questions. This is not surprising, given that this is 'cutting edge research'. Furthermore, basic research and adaptive research were being conducted in parallel; more than this, however, the traditional separation between the two was becoming increasingly blurred, with adaptive research expected to shed light on some of the 'surprising' outcomes of basic research. This complex reality of multiple uncertainties and a blurring of traditional boundaries between upstream and downstream activities can be contrasted with a consistent framing of the enterprise by its main promoters as managing – against considerable odds – to follow a classic linear innovation model (Rogers, 2003). At every stage, technical uncertainties were shielded from debate – both within and beyond relevant networks. These uncertainties remained, however, and continued to accompany the Golden Rice materials as they travelled downstream.

In the process, Potrykus's assertion in 2001 that the scientific problems were 'overcome' (Potrykus, 2001, p1159), implying that the subsequent technology transfer would be unproblematic, remained the overarching frame for the Golden Rice project. Questions of pathways not taken – for example the relative merits of continuing with the prototype (for which introgression work was well under way), or even eschewing both the prototype and the more recent Syngenta events in favour of generating starting materials *in situ*[42] and thus free of TP constraints – remained unexplored. This created a particular set of tensions within a project that, from its premature birth as a science policy controversy, had increasingly institutionalized a defensive position. With the IP and scientific problems 'solved', the Humanitarian Board focused on anticipated regulatory hurdles and consumer resistance as the obstacles to Golden Rice dissemination,[43] an assertion that chimed with its earlier problematization of debates around Golden Rice as a contest between promoters and an anti-GM or even 'anti-science' lobby (Potrykus, 2001, p1161).

Framing 'Acceptance': The Case of the Philippines

These dynamics came sharply into focus in the case of the Philippines. Both the state and civil society have had a particular relationship with the IRRI, an international institution located on its soil, beginning with a historic convergence in 1965 when an election campaign using the slogan 'progress is a grain of rice' brought Marcos to power, accelerating the release of IR-8 as 'miracle rice' (Cullather, 2004, p243). Since the EDSA[44] revolution of 1987, however, relations between IRRI and Philippine civil society have often been conflict-ridden, to the point that 'civil society' has become a shorthand term used informally at IRRI to refer to opposition, particularly towards transgenic research and especially towards Golden Rice.

One key point of conflict between IRRI and Filipino NGOs, since the late 1980s, has been the issue of biosafety. Largely as a result of this relationship, the Philippines was one of the early countries to develop a national biosafety framework, long before it became an issue on the international stage. As one activist recalls:

> *It was a big issue, and a real one because IRRI is here* ... Biosafety [has been] here since 1989 ... because of their [IRRI's] earlier research on tungro virus and blast, [we] had no biosafety regulations then ... so you had an institute with the capacity and competence to do it ... in a country where the regulations were not ready. [This was] raised in Congress ... congressional hearings in 1987–88 ... IRRI nearly folded ... had to give in. [This was] at a time when the relevance of IRRI was on the table ... NGOs were questioning ... [This was at] the height of EDSA [when there were] sweeping reforms ... [The Philippines' biosafety regulatory system established under Executive Order 430 was] hailed as the only, the first biosafety regulations in Asia ... long before Cartagena.[45]

These events set the tone for an uneasy relationship between IRRI and the sustainable development movements and NGOs that proliferated during the 1990s, particularly after the UNCED Conference in Rio of 1992 – including Southeast Asia Regional Initiatives for Community Empowerment (SEARICE), Magasaka at Siyentipiko Para sa Pag-unlad ng Akricultura (MASIPAG – Farmer-Scientist Partnership for Development), and the Philippine Greens, among others. From 2002, however, events changed course quite dramatically when the regulation of the commercial release of transgenic crops shifted from the Department of Science and Technology (DOST), the home of the National Biosafety Committee, to the Department of Agriculture (DA). According to NGO observers, in 2002 the DA 'pulled the rug from under DOST',[46] issuing Administrative Order 8 on the 'Impact and Commercialization of GM crops'.[47]

In December 2002, the government approved the release of Monsanto's *Bacillus thuringiensis* maize – '*Bt* corn' as it is known in the Philippines – a maize variety genetically modified to be resistant to the corn borer pest.[48] This decision is noteworthy in that, unlike other countries in the region, the first GM crop to be released in the Philippines was a *food* crop.[49] As one Filipino biotechnology advocate commented; 'we were brave, others [commercialize *Bt*] cotton; here we are with corn ... you can eat it'.[50]

The sustainable development NGOs, however, were aghast both at the decision and the speed with which it was reached and enacted. One activist recalls 'a feeling that the Philippines was being left behind by its neighbours ... *as if we are in a race*'.[51] In May 2003, a group of NGOs began a hunger strike outside the DA, demanding a moratorium on the commercialization of *Bt* corn, subject to further testing. In retrospect, the strike, which ended in failure after 30 days, is regarded as a turning point in the fortunes of the anti-GM movement. *Bt* corn received final authorization in December 2003 and by this time the anti-GM movement had splintered; as one activist observed 'we are still in the aftermath of that'.[52]

A common perception held by IRRI and PhilRice scientists, as well as some nutritionists in the Philippines, is that there is strong opposition to GM crops in the country and this is largely the work of Greenpeace, which in turn is driven by European funds and agendas. While this framing of GM opposition as a 'downloaded controversy'[53] may be justifiable in some national contexts, in the Philippines this is far from the case. However, the hiatus following the hunger strike and the arrival, in 2002, of Greenpeace, with their characteristically high-profile approach, made it appear so. As a consequence, the heightened nature of these debates in recent years has tended to reinforce a notion of the policy context for Golden Rice as an obstacle course constructed by 'civil society' to resist its implementation.[54] This polarized interpretation of a complex policy landscape is reproducing processes similar to those that took place when the Golden Rice controversy erupted on the international stage.

Field testing began in the Philippines in April 2008, with commercial release envisaged in 2011,[55] although this timetable represented one of a series of postponements (Al-Babili and Beyer, 2005). At this point, questions still remained around the bioavailability and post-harvest stability of the beta-carotene content of Golden Rice (though studies were by then under way in the United States and China which, it was hoped, would shed light on some of these questions).[56] Nevertheless, uncertainties acknowledged by IRRI scientists in 2001, when the materials were first transferred to IRRI, remained largely unanswered:

> The international research institute (IRRI) took up this challenge in January 2001 and has succeeded in transferring the provitamin A trait from the modified *japonica* rice to an *indica* background (which is more suitable for cultivation in Asia) ... however, doubts remain over the nutrient's bioavailability – the extent to which the added nutrient is taken up by the body of the person consuming it – in the edible part of the plant ... Questions have

also been asked about the stability of the provitamin A during storage and post-harvest processing, as well as its thermal stability during cooking.[57]

The question of acceptance of Golden Rice has been framed primarily in terms of its acceptance by farmers as a GM rice variety. In 2003, findings of PhD research to 'survey the risk perceptions of golden rice among farmers in Nueva Ecija' were quoted in Philippine publications under the headline 'NE farmers see no risk in Golden Rice' (PhilRice, 2003, p21; Zanago, 2003, p32). But *which* rice would be golden? Ultimately farmers' decisions will reflect their assessment of the varieties selected for introgression, as well as the 'golden' trait itself. Scientists at IRRI have chosen to backcross Golden Rice with IR64, its flagship 'premium' variety, which raises questions about its 'pro-poor' focus.[58] Similarly, PhilRice chose PSB Rc 82 and 128, both in a similar category.[59] The question of varietal selection, therefore, was not put to farmers, poor or otherwise.

Similarly, consumer acceptance has been implicitly linked to a perceived reduction in public concern following the commercialization of *Bt* corn.[60] However, it is well known that an everyday meaning of yellow rice indicates that it has been stored for too long.[61] Furthermore, the impact of the 'golden' trait on the eating quality of the final product remains an open question. Assumptions that Golden Rice will be accepted across Asia as a substitute for saffron rice,[62] a popular dish in parts of South Asia, reveals an ignorance of the diversity of rice cultures across the continent. However, qualities of taste and texture are not so easily dismissed, given the importance of consumer preference in shaping rice markets.[63] Cereal chemists draw attention to the possible effects of increased beta-carotene on grain quality, noting that it has been 'difficult to get enough rice' to test eating quality. Moreover it was 'difficult with early backcrosses, what to compare ... how much *Cocodrie* is still in it? However, from the way the rice looks, there are gross conclusions you can make', that eating quality will be affected. At the same time, as with orange sweet potato, the texture of a rice variety would be expected to change with the introduction of the golden trait since 'it's a direct response to beta-carotene'.[64]

On the Golden Rice website is the assertion that 'It's just rice'.[65] Reviews of the place of rice in the social, cultural and spiritual fabric of the Philippines (Asia Rice Foundation, 2004; Castillo, 2006), however, reveal that it is *never* 'just rice'. In this case, the framing of 'consumer acceptance' as the last in a sequence of 'roadblocks' standing in the way of the Golden Rice project fails to recognize myriad ways in which 'rice is intimately woven into the warp and woof of Filipino traditional life' (Hornedo, 2004, p5):

> Rice is more than a 'grain of rice'. It is arts and culture; history; politics; tradition; metaphor; land and labour relations; increasing rural-urban connectedness; repeated promises of self sufficiency; keeping a World Heritage site [the Ifugao rice terraces, Northern Luzon, the Philippines]; farmers in transition;

national sovereignty vs. reciprocity ... As Doreen Fernandez put it, 'If we did not have rice, our deepest comfort food, we would probably feel less Filipino'. (Castillo, 2006, pxi)

These events linking Golden Rice, IRRI and its host country are illustrative of the transfer of not only the Golden Rice materials, but of a bundle of assumptions and expectations that, once the hurdles are surmounted, Golden Rice would be absorbed into national programmes as a nutritional intervention, and accepted by poor farmers and consumers as a commercially viable crop and staple food. These assumptions have been implicit throughout, despite the reframing of the initial research as merely a 'proof of concept' study. In this case, the familiar notion of a linear innovation path provided refuge from a host of uncertainties that continued to throw doubt on many aspects of the project, providing promoters with a platform from which to project the responsibility for the controversies that these questions inevitably raised back on to an 'irrational' opposition.

Conclusion

The series of events through which Golden Rice was transformed from a emblem of international public research collaboration to an exemplar of 'a new type of public-private partnership', is illustrative of shifts that took place in the organization and funding of international research in the intervening years. A thread of continuity in these evolving arrangements, however, was a conventional notion of development as a one-way technology transfer. In the context of current intellectual property regimes, both public- and private-sector actors in the North appear to interpret their patent-releasing function in terms of a 'white man's burden' to share the fruits of advanced technology with partners in the South. In this context, the case of Golden Rice highlights a new consilience between enduring 'top-down' development models and evolving North-South power-knowledge relations.

At the same time, this chapter has revealed a mismatch between an image of 'cutting-edge research' and a reality of ongoing processes of enquiry generating uncertain and unresolved outcomes. In particular, an initial emphasis on the need for new types of institutional arrangement to deliver an imagined finished product, together with the creative use of 'proof of concept' as a boundary term, shaped and often curtailed discussion of emerging scientific questions, pushing these further downstream to be absorbed into later adaptive breeding and implementation stages, with uncertain results. These developments were overseen by an initially close-knit, but increasingly dispersed, Humanitarian Board that, crucially, lacks the grounded perspective of the iron rice network which, as discussed in the previous chapter, was so firmly located in the Philippine context.

This is, therefore, linear innovation with a twist; a twist in which existing concerns about adoption, ownership and impact are compounded by new scientific and policy uncertainties. In this case, the lack of attention to downstream complexity and context-responsiveness was exacerbated by an oversimplification of the nature and outputs of upstream research and the black-boxing of choices made and questions still unresolved. In the case of Golden Rice, a felt need to defend the project against opposition has intensified this oversimplification, when the project would have benefited from a more open discussion. In the context of contemporary discourses advocating a 'freedom to innovate' (Juma and Serageldin, 2007) as the key to reviving agricultural development, the frequency with which Golden Rice is cited as a case of a proven technology blocked by irresponsible regulators and activists (for example, see Taverne, 2007) raises serious concerns.

While discussion of these outstanding uncertainties was sidelined, international attention was again refocused upstream. Under the Gates Foundation's Global Health programme, an ambitious research programme and international consortium[66] had come to together around 'engineering rice for high beta-carotene, vitamin E, protein and enhanced iron and zinc bioavailability'. Research was proceeding in several locations on the assumption that other traits will be stacked into the completed – but still untested – Golden Rice product. While conducted by a diverse set of institutions, the governance structure includes familiar names, with a steering committee including Beyer, Barry and Dubock, overseen by the Humanitarian Board acting as 'an external advisory board'.[67] These developments suggest that, far from engaging in a grounded discussion about the 'real' obstacles to delivery and impact, attention and funding shifted again to increased upstream technical and institutional complexity (in a similar way that uncertainties around iron rice research have been avoided with the arrival of HarvestPlus, as will be seen in the next chapter).

Placing Golden Rice at the centre of this ambitious, large-scale research programme took as given that: first, vitamin A is and remains a significant public health issue in target countries; second, the finished Golden Rice product will be sufficiently bioavailable and stable; and, finally, Golden Rice will be accepted by farmers and consumers (whichever varieties are selected for introgression), once resistance to GM crops has been eroded. Rather than opening up debate around these assumptions, such developments raise the stakes yet higher and in the process close off further inquiry and discourage open communication and discussion of findings.

4

An Alliance Around an Idea: The Shifting Boundaries of HarvestPlus

Introduction

The last two chapters traced early biofortification pathways that emerged in the 1990s. While they were products of different institutional histories and areas of scientific enquiry, these early initiatives shared certain key characteristics. The first of these could be called the missionary effect: in each case these projects relied on the determination and vision of key individuals and the small, close-knit actor-networks that formed around them. Second, each project was accompanied, in its early stages, by modest expectations of success; these projects were exploratory in nature. In each case, however, as time progressed, the projects underwent gradual metamorphoses from an open, exploratory mode, through successive stages of institutional framing, 'black-boxing' and simplification, towards linear approaches that have effectively discouraged scientists from 'looking sideways',[1] to consider the multiple uncertainties emerging from this new science.

This chapter explores the processes by which these dynamics intensified once these initiatives were absorbed under the umbrella of HarvestPlus, one of the programmes selected by the CGIAR in the early 2000s to pilot a new approach to the conduct and funding of research, called the 'Challenge Program'. In accumulating these various projects, each with their own historical pathways, played out in a variety of locations, HarvestPlus was ambitious in scope, extending the actor-networks supporting biofortification research, now a global project, in many directions. Simultaneously a programme of international research and an alliance around an idea,[2] HarvestPlus appears as an exemplar of the type of interdisciplinary, multi-institutional collaboration envisaged for the CGIAR, designed to shift the role of CGIAR centres from research institutions to 'brokers' of global networks that can generate research outputs as international public goods (IPGs) amenable to adaptation and adoption worldwide.

How did previously fragmented efforts – including iron rice research at IRRI and Potrykus and Beyer's Golden Rice project – come to congregate under the HarvestPlus umbrella? While retaining as director Howarth 'Howdy' Bouis, who had managed the predecessor CGIAR micronutrients initiative from which iron rice study emerged as the flagship project (see Chapter 2), HarvestPlus was a radical departure from earlier projects in nature, scale and scope. Chapter 1 highlighted some of the broader trends in international development and nutrition that have emphasized cross-sectoral synergies and goal-oriented, micronutrient-based, nutrition programming. This chapter follows the processes through which, encouraged by these trends, biofortification moved from the outer margins of international crop research to attract support from the newly instituted CGIAR Science Council and the Bill and Melinda Gates Foundation, now one of the world's largest private philanthropic organizations.

As an alliance formed around an appealing but still largely untested idea, the HarvestPlus network appeared as fragile as it was far-reaching. In contrast, the iron rice 'family', discussed in Chapter 2, though interdisciplinary and multi-institutional, was clearly located and grounded in a Southeast Asian, Philippine context; the Golden Rice Humanitarian Board and Network, discussed in Chapter 3, while international, have been characterized by a close-knit core managing access to knowledge and materials through institutional arrangements emphasizing vertical over lateral relations. This chapter traces events unfolding during the early years of HarvestPlus in an attempt to reveal how and why, despite its faltering progress as an research and development programme, HarvestPlus has endured as the international biofortification 'mother ship'[3] and platform for spin-off initiatives of various kinds.

Back to Basics? A Challenge Program

While the CGIAR micronutrients project (1994–9) had been a relatively modest programme, it carried a set of expectations that it would make the case for a larger scale initiative. The IRRI-hosted 'Improving Human Nutrition through Agriculture' seminar[4] in 1999 had provided a platform to share findings and build broader support; however, initially, this only extended to ADB support for the high-iron rice project. In the meantime, Howdy Bouis continued to seek support for a broader, multiple crop initiative within the CGIAR, submitting a proposal for an IFPRI-led 'system-wide programme' to the CGIAR Technical Advisory Committee (TAC). The TAC rejected this proposal on the basis that biofortification was 'not a priority area for the CGIAR'.[5]

As discussed in Chapter 1, four years later a revised proposal, now jointly sponsored by IFPRI and CIAT, was approved by the CGIAR Interim Science Council, the body that had superseded the TAC as a result of organizational reforms under way in the CGIAR. What would account for this shift in position? While the Science Council was a new structure, at this point in the transition process the Interim Science Council was composed of the same people as the TAC that had initially rejected the proposal. It is instructive, therefore, to consider the combination of factors that could have led this group of people, four years later, to reach a very different decision.

The replacement of the TAC, 'a broad mix of people with research and development background', with the Science Council, 'consisting of a few, high level science policy strategists' each recognized for their 'solid scientific stature' (CGIAR, 2001, pp1, 6), was one of the key elements of the CGIAR reform programme. This organizational change was indicative of a more fundamental shift of the CGIAR 'back to its roots' as a research institution.[6] While reflecting donor concerns about development impact the Science Council mandate embodied a re-emphasis on 'Research for development – not development *per se*' (Science Council, 2006, p7).

This new body initiated a process of 'system-level priority setting' that would guide CGIAR research for the period 2005–15 according to three criteria:

- the expected *impact* on poverty alleviation, food security and nutrition, and sustainable management of natural resources, taking into account the expected probability of success and expected impact if successful;
- the degree to which the research provides *international public goods*; and the existence of alternative sources of supply of the research; and
- the CGIAR's *comparative advantage* in undertaking the research (Science Council, 2006, pp5–6, emphasis added).

This process generated '20 research priorities for the CGIAR, organized within five priority areas' (Science Council, 2006, p6) (see Box 4.1). These included, notably, 'Priority 2C: Enhancing nutritional quality and safety'. Anticipated direct and indirect impacts of these research priorities on achievement of the MDGs were set out in some detail (Science Council, 2006, pp5–7).

The Science Council proposed that, after a period of transition, CGIAR members and centres should allocate '80% of the total CGIAR budget for research and related capacity strengthening' to the priority areas identified (Science Council, 2006, p6). Throughout its summary report, the Science Council reasserts the role of the CGIAR in generating research for development as 'international public goods'. The proposed shift 'away from development activities with no research content' was endorsed on that basis: 'The SC [Science Council] is confident that a stricter application of the criteria to sharpen the scope of research will open up new opportunities for longer-term impact through strategic research activities' (Science Council, 2006, p7).

A second element of the CGIAR reform process was 'the development of *Challenge Programs* that respond directly to major concerns on the global development agenda' (CGIAR, 2001, p1, original emphasis). 'The CP [Challenge Program] became one of the four pillars of the CGIAR reform

Box 4.1 CGIAR System Priorities, 2005–15

Priority area 1: Sustaining biodiversity for current and future generations

Priority 1A: Promoting conservation and characterization of staple crops

Priority 1B: Promoting conservation and characterization of underutilized plant genetic resources

Priority 1C: Promoting conservation of indigenous livestock

Priority 1D: Promoting conservation of aquatic animal genetic resources

Priority area 2: Producing more and better food at lower cost through genetic improvements

Priority 2A: Maintaining and enhancing yields and yield potential of food staples

Priority 2B: Improving tolerance to selected abiotic stresses

Priority 2C: Enhancing nutritional quality and safety

Priority 2D: Genetically enhancing selected high-value species

Priority area 3: Reducing rural poverty through agricultural diversification and emerging opportunities for high-value commodities and products

Priority 3A: Increasing income from fruit and vegetables

Priority 3B: Increasing income from livestock

Priority 3C: Enhancing income through increased productivity of fisheries and aquaculture

Priority 3D: Promoting sustainable income generation from forests and trees

Priority area 4: Promoting poverty alleviation and sustainable management of water, land, and forest resources

Priority 4A: Promoting integrated land, water and forest management at landscape level

Priority 4B: Sustaining and managing aquatic ecosystems for food and livelihoods

Priority 4C: Improving water productivity

Priority 4D: Promoting sustainable agro-ecological intensification in low- and high-potential areas

Priority area 5: Improving policies and facilitating institutional innovation to support sustainable reduction of poverty and hunger

Priority 5A: Improving science and technology policies and institutions

Priority 5B: Making international and domestic markets work for the poor

Priority 5C: Improving rural institutions and their governance

Priority 5D: Improving research and development options to reduce rural poverty and vulnerability

Source: Science Council, 2006, p4.

programme' (Science Council and CGIAR Secretariat, 2004, p4): 'A CP is defined as: "*A time-bound, independently-governed program of high impact research, that targets the CGIAR goals in relation to complex issues of overwhelming global and/or regional significance, and requires partnerships between a wide range of institutions in order to deliver its products*"' (Science Council and CGIAR Secretariat, 2004, p4, original emphasis).

In 2001, ten proposals were identified as 'candidates for acceleration' as Challenge Programs, including a project entitled: 'Harnessing Agricultural Technology to Improve the Health of the Poor: Biofortified Crops to Combat Micronutrient Deficiency' (CGIAR, 2001, p6). At the CGIAR AGM that year, the decision was taken 'to accelerate the process with the launch of three CPs [Challenge Programs] on a pilot basis'. In 2003, three Challenge Programs – Water and Food, 'HarvestPlus' (formerly the Biofortification CP) and Generation (formerly the Genetic Diversity CP) – were launched (Science Council and CGIAR Secretariat, 2004, p4).

In 2004 the Science Council and CGIAR Secretariat synthesized lessons to date from implementing the three pilot Challenge Programs. This synthesis document (Science Council and CGIAR Secretariat, 2004) is instructive in that it provides an articulation of the principles behind the Challenge Program mechanism and an indication of how these principles are to be applied in practice. A central theme is the importance of generating 'time bound outputs of IPG nature'. This is presented in terms of an international public goods model understood as 'the comparative advantage of the CGIAR system': 'The CGIAR has potential to seek major efficiency in the application of basic science to solve similar problems in multiple domains. This can be called the "comparative paradigm", *which is precisely what the CPs are about* ... The Generation CP and HarvestPlus CP are seeking proof of concept of this "comparative paradigm" which holds promise for widespread impact' (Science Council and CGIAR Secretariat, 2004, p7, emphasis added).

Complementing the international public goods model as the thread of continuity and the CGIAR's *raison d'être* was the principle that added value can be leveraged through engagement in strategic partnerships. This theme was further developed in the 'Business Plan' prepared for the Water and Food Challenge Program. This document outlines a 'business model' characterized by a consortium approach and open competitive grants (Rijsberman, 2002, p2). In this case the role of CGIAR centres shifts to the role of 'broker':

> In this changing world the role of Future Harvest centers[7] changes from international research organizations that initiate and have primary responsibility for doing research in the developing world, *to organizations that derive their added value primarily from brokering and facilitating international research networks*. The international research centres link ARIs[8] and NARES[9] in complex multi-disciplinary research programmes with a strong focus on poverty alleviation and capacity building. The brokering role is a

substantive role that *does also require the maintenance of high quality research capacity within the system of international centres*. The nature of the role of the Future Harvest centers should, however, adapt itself to playing different roles: from (1) providing a 2-way international window on the world for large, high capacity countries such as Brazil, India or China, to (2) playing a major role in building capacity for research in countries with severely restricted internal capacities. (Rijsberman, 2002, p3, emphasis added)

This 'strategic partnership' theme within the Challenge Program design dovetailed with ongoing debates within the CGIAR regarding engagement with the private sector. In 2005 IFPRI convened an international dialogue on 'Pro-Poor Public-Private Partnerships on Food and Agriculture' (IFPRI, 2005). This was in a context of concerns about the CGIAR's redefinition from a public research body to 'a strategic alliance of 63 countries, international and regional organizations, private foundations supporting international agricultural research centres that work with national agricultural research systems, the private sector and civil society'. These concerns came to a head in 2002 when the Syngenta Foundation was appointed to the CGIAR board, prompting the CGIAR NGO Committee to 'freeze its membership'.[10]

While acknowledging that public-private partnerships 'are not a panacea for all development challenges in agriculture', IFPRI reported 'a broad consensus that [public-private] partnerships can create valuable synergies through knowledge sharing, joint learning, scale economies, resource pooling and risk sharing' (IFPRI, 2005, p4). Notably, the Challenge Program mechanism was highlighted as presenting a way forward:

Partnerships with CGIAR Centres require that the CGIAR System rethink its structure and leadership. The longstanding principles of decentralization and centre autonomy are not helpful in dealing with the private sector. The system's recent creation of Challenge Programs – large, multi-stakeholder partnerships focused on major global issues, such as water for food and breeding micronutrient-dense staple crops – offer a much better model from the private sector's point of view. (IFPRI, 2005, p6)

These shifting priorities transformed the fortunes of Bouis's biofortification initiative from an interdisciplinary project 'falling between the cracks' in the CGIAR system to an exemplar of the potential offered by a new type of strategic research collaboration envisaged for a repositioned CGIAR. In 2003 the Biofortification Challenge Program, rebranded HarvestPlus, was selected as one of the three pilot Challenge Programs, which would be organized as follows:

> Activities will be undertaken by an international alliance of Future Harvest Centers, national agricultural research and extension systems (NARES), departments of human nutrition and plant science at universities in developing and developed countries, advanced research institutes (ARIs) with expertise in micronutrients in plants and animals, and genomics, nongovernmental organizations (NGOs), farmers' organizations in developing countries, and private-sector partnerships. The Future Harvest Centers involved in the Biofortification Challenge Program are world renowned for their plant breeding expertise and extensive germplasm banks, strong ties to national agricultural extension programs, and links to the human nutrition community. Thus, they are well placed to coordinate the proposed activities. However, close collaboration with institutions that offer complementary scientific expertise, skills, and experience not found within the Future Harvest Centers, is critical to a successful outcome. To achieve the goals and objectives of the Program, new ways of working together, both within the CGIAR system and with external partners, are needed. (CIAT and IFPRI, 2002, piii).

The programme was immediately awarded US$3million in World Bank funding, under an existing agreement with the CGIAR (as part of its support for the CGIAR reform programme) to fund all the successful Challenge Program candidates.

One observer has remarked that, of all the Challenge Programs, HarvestPlus is the easiest to remember.[11] This may be because it incorporated certain key characteristics of the Challenge Program model, facilitating the enrolment of the Science Council, in a way that had not been possible in the era of the TAC. First, it addressed a problem that was high on the global development agenda: as discussed in Chapter 1, large-scale micronutrient interventions rate highly in terms of cost-effectiveness in an era of MDG-driven development.[12] Second, it addressed a problem that is complex – and therefore required, justified even, the type of heterogeneous networks and complex structures envisaged for the newly conceived Challenge Program. Crucially, such a structure presented a way forward for the CGIAR to maintain a central, if transformed, role in international crop research.

This ability of the *idea* of HarvestPlus to hold together such levels of complexity and heterogeneity, while packaging it in terms of a relatively straightforward, memorable formula, has been key to the transformation of biofortification research into the 'global' project it has become. At the same time, this repackaging resonated with an increasingly hegemonic, overarching frame for global development; a framing that has been consolidated by the arrival on the scene of a major new development actor, and the largest donor of HarvestPlus, the Bill and Melinda Gates Foundation.

A Turning Point: Enrolling the Gates Foundation

In 2001 Bouis had approached the Bill and Melinda Gates Foundation (BMGF) about funding a larger scale, multi-crop biofortification programme. At this point the foundation was relatively young, with a staff of five to six people.[13] The initial assessment was unfavourable, particularly given that, at that stage, Bouis had not secured support from any major donors for such an expanded programme. In the following year, however, a series of serendipitous personal connections occurred. Sally Stansfield of the BMGF was visiting the Centro Internacional de Agricultura Tropical (CIAT – International Center for Tropical Agriculture) to discuss the possibility of funding biofortification research on the common bean, on a single crop basis. Stansfield and Joachim Voss, then director-general of CIAT, were former colleagues at the International Development Research Centre (IDRC) in Canada.[14] Voss's decision to invite Bouis to attend the meeting to explore the possibility of extending the proposal to a multiple crop project transformed Bouis's project from an IFPRI initiative to a joint IFPRI/CIAT proposal with the full support of a CGIAR crop-breeding centre.[15]

Bouis approached the BMGF for a second time in 2003 with an altogether different result. By then, the programme had been approved by the CGIAR Interim Science Council for acceleration as a pilot Challenge Program. As such, the proposal had been assessed through the Challenge Program peer review process and allocated funding from the World Bank and other sources. The BMGF was therefore presented, crucially, with a co-funding proposition and with a project already approved through what BMGF staff regarded as a sufficiently rigorous assessment process: 'That the World Bank funding was already in place helped in discussions with the Gates Foundation. The eight anonymous, external reviews commissioned by the iSC [Interim Science Council] were made available to the Gates Foundation and this shortened the time required in their review process' (HarvestPlus, 2004b, p4).

In 2003 the BMGF approved US$25 million over four years (IFPRI, 2003). At this point, the primary donors of HarvestPlus were the World Bank (US$3 million per year), the BMGF (US$6.25 million per year) and USAID (US$2 million per year) (BCP, 2003, p1).[16] This funding was for the first phase (2004–7), which would focus on six staple crops (rice, maize, wheat, cassava, sweet potato and common bean) and three micronutrients (vitamin A, iron and zinc). The following excerpt from the programme proposal outlines a cautious approach to the use of more controversial technologies; the first phase would concentrate on the potential of conventional plant-breeding methods, with transgenic research (and additional crops) written into the second phase:

> The Biofortification Program will focus on three micronutrients that are widely recognized by the World Health Organization (WHO) as limiting: iron, zinc, and vitamin A. Full-time breeding programs are proposed for six staple foods for which feasibility

studies have already been completed and which are consumed by the majority of the world's poor in Africa, Asia, and Latin America: rice, wheat, maize, cassava, sweet potatoes, and common beans. Pre-breeding feasibility studies are proposed for eleven additional staples: bananas, barley, cowpeas, groundnuts, lentils, millet, pigeon peas, plantains, potatoes, sorghum, and yams. Breeding, dissemination, and impact activities, outlined in the ten-year plan, are focused on development of conventionally-bred crops. No activities involving the release of nutritionally-improved transgenic crops to farmers and consumers are proposed here or are included in proposal budgets for the initial four years for which funding is being requested. Research and development activities with respect to transgenic crops are confined to agricultural research centers and research laboratories. Transgenic methods hold great promise for improving the nutrient content of staple foods and speeding up the breeding process over what can be achieved using conventional methods. High social benefit and lower risk applications, such as the incorporation of desirable traits from crop wild relatives, will be favored throughout the program whenever transgenic methods are considered. (CIAT and IFPRI, 2002, piv)

'A new class of philanthropist'

Why was the BMGF prepared to risk such substantial funds on an as yet unproven approach? Foundation staff have referred to the 'cogent argument' presented by a proposal with potential to deliver 'attributable impact'.[17] However, it is important to place the foundation's support for this initiative within the broader context of its prioritization of 'Global Health' (Gates Foundation, 2000, p3), and, within that, for its ongoing support for conventional food fortification projects.

The BMGF had, since its launch in 1999, supported two US NGOs pioneering technology transfer of conventional food fortification technologies (Gates Foundation, 1999; Gates Foundation, 2001): the Program for Appropriate Technology in Health (PATH), known for its 'Ultra Rice' technology (PATH, 2005; PATH, 2006), and Sharing US Technology to Aid in the Improvement of Nutrition (SUSTAIN), known for its work on iron powder for fortification and for advocating the fortification of US (PL480) food aid.[18] Then, in 2002, the foundation 'announced a $50 million commitment to support the formation of the Global Alliance for Improved Nutrition (GAIN) to reduce vitamin and mineral deficiencies among children in developing nations' (Gates Foundation, 2002, p13). This alliance would work towards establishing national level public-private partnerships for food fortification in GAIN's member countries (GAIN, 2005).

A decision by the BMGF to co-fund HarvestPlus can be viewed as part of a broader investment strategy in which the foundation was spreading its risk.

In 2005, the director of Rockefeller Foundation, Judith Rodin, drew attention to 'a new generation of business-minded philanthropists, led by Bill Gates'. Rodin's vision of a new Rockefeller Foundation acknowledges the influence of 'the new philanthropists' who emphasize 'the importance of being "strategic"; of leveraging the relatively small sums of money at its disposal ... through partnerships; and, above all, of achieving "impact"'.[19] Notably, these elements of a more business-minded approach to philanthropy were mirrored in changes then under way in the CGIAR and exemplified by the vision underpinning the Challenge Program design. In a keynote speech at the 2007 'Global Philanthropy Forum', hosted by Google, Rodin argued that the US philanthropy sector 'must embrace Silicon Valley-style, market-based approaches'.[20] Another speaker noted that: 'Philanthropy has always had innovation in its DNA. I see that is being ratcheted up ... What was already an inventive sector now has an infusion of talent of young, energetic, socially conscious people who have gained a new wealth, a new class of philanthropist.'[21]

This reference to a new class of philanthropist has also been endorsed by Jeffrey Sachs of the Millennium Project, who believes that these 'wealthy philanthropists' have the potential to 'eclipse the G8' in addressing global development problems: 'The Rockefeller Foundation was the world's most important development organization of the 20th century, and the Gates Foundation can be that of the 21st century ... Gates can make a huge difference if they hit the right model.'[22]

Announcements such as these convey high expectations of a new generation of philanthropists, who made their fortunes in the US-based high-tech industries, to succeed where others have failed in tackling enduring development problems as long as they 'hit the right model'. Who are these individuals and why do they inspire such confidence?

The BMGF is the largest private foundation in the US and considered the leader of this new phenomenon. It was formed in 1999 from the consolidation of the William H. Gates Foundation (founded by Bill Gates Sr) and the Gates Learning Foundation (founded by Bill Gates Jr and his wife, Melinda French Gates) (Gates Foundation, 1999, p3). By January 2005, Bill and Melinda Gates had 'endowed [the] foundation with more than $28.8 billion ... to support philanthropic initiatives in the areas of global health and learning'.[23]

William (Bill) H. Gates Jr was co-founder and former chairman of the Microsoft Corporation, where he became 'known for his aggressive business tactics and confrontational style of management'.[24] Melinda Gates had worked as a project manager at Microsoft from 1987 to 1996.[25] In 1994 she married Bill Gates and in 1996 she left Microsoft, after the birth of their first child. One commentator has suggested that 'her mix of grace and gravitas has tempered the brash image of her tech-tycoon husband, and has eased acceptance of their ambitious agenda for ending health inequities'.[26]

In 2006, Warren Buffett, Bill Gates's 'friend and bridge partner', set a new precedent by announcing that he would donate his fortune to the BMGF – and not to his own foundation – doubling its budget overnight (Okie, 2006, p1086). The then BMGF co-chair, Patty Stonesifer, wrote in that year's annual report:

> On June 26, Warren Buffett announced an astonishing pledge to the foundation – 10 million shares of Berkshire Hathaway Inc. stock worth more than $31 billion at the time ... Giving away money isn't hard. But giving it away effectively sure is. We were already making about $1.4 billion a year in grants. Last year, the total jumped to $1.6 billion, thanks to Warren's pledge, and we'll be giving away a projected $3.2 billion a year by 2009. We're not making the jump all at once; our annual grant making will increase by about one-third in each of the next three years. (Gates Foundation, 2006, p3)

Known as 'the Gatekeeper', Stonesifer had been a key member of the BMGF team since the early days, bringing 'a new style of leadership to philanthropy'.[27] Her biography clearly shows her membership of the 'new class of philanthropist':

> In 1997, at the age of forty, Patty Stonesifer left her executive position at Microsoft for early retirement as a multi-millionaire. While Stonesifer was looking forward to spending time with her two teenage sons and working as a part-time consultant for DreamWorks SKG, Bill and Melinda Gates were working on a project to provide donated computers to public libraries in poor neighborhoods. The Gateses invited Stonesifer to tour several libraries that would benefit from the initiative, and, feeling obliged, the former 'Microsoftie' accepted. It took just one trip to a small town in South Dakota, where the local Rotary had pooled their money to buy the library a single computer, for Stonesifer to agree to head up the Gateses' library project. With no previous professional philanthropic experience beyond her own million-dollar donation in 1998 to the Multi-Service Centers, a Redmond, Washington-based family crisis program to which she had volunteered a considerable amount of time, Stonesifer has become a central figure in the Gateses' philanthropic work.[28]

Despite a relative lack of previous 'professional philanthropic experience', Stonesifer had clear ideas about how an endowment such as that of the Gates can be mobilized to make an impact:

> Stonesifer points out that although Gates's assets are vast, a gift of just $350 each to every American would extinguish his fortune. *'You can do a lot with the money',* Stonesifer says, *'or*

you can dribble it away. One reason why Bill and Melinda are committed to giving it back is that it makes most sense when you divide it into what it can do. As a giant bucket, it's kind of an irrelevant number'.[29]

While presenting a break with the past, however, these new philanthropists have retained one key characteristic from the previous generation: a belief in the potential of technological solutions to provide lasting solutions to intractable, complex problems. A quote highlighted in the BMGF's 2002 *Annual Report* is illustrative: 'For a long time, I've had a love for how science and technology can be integrated with public policy to solve unbelievably difficult and important problems facing the human condition' (Rick Klausner, Executive Director, Global Health, quoted in Gates Foundation, 2002, p12).

Similarly, a 'letter from Bill and Melinda Gates' posted on the BMGF website states: 'We also believe in the power of science and technology to improve people's lives. In recent years, the world has made tremendous advances in fields ranging from biology to information technology, and yet not everybody is benefiting from these innovations. Our goal is to help apply science and technology to the problems of the neediest people.'[30]

This belief in the potential of science and technology, combined with a new 'business-minded' approach, found its clearest expression in the foundation's 'open innovation' model. In 2003 the BMGF launched its 'Grand Challenges for Global Health', inviting grant applicants to participate in a process of open competition. One of the successful bidders was an international research consortium proposing to conduct a programme of biofortification research that promised to be more ambitious and high-risk than HarvestPlus:

Launched in 2003, the initiative unfolded in two stages. First, an international scientific board issued a call for ideas: What scientific and technological innovations, it asked, could have the greatest impact on health in the developing world? After reviewing more than 1,000 ideas, the board identified 14 Grand Challenges that, if solved, could save millions of lives in developing countries. These challenges include, for example, vaccines that don't require refrigeration, vitamin-fortified staple foods, and more effective and easy-to-use diagnostic tools. In the second stage, the board issued a call for research proposals based on the 14 Grand Challenges, and scientists from 75 countries submitted more than 1,500 funding requests. In June, the Grand Challenges initiative announced grants totalling $436.6 million to support 43 projects. One would develop a chemical to prevent mosquitoes from smelling humans, which could stop them from being able to transmit disease. Another would design a hand-held diagnostic device that could be used in developing countries to test a drop of blood for a battery of diseases. Four others would develop new varieties of cassava, rice, sorghum, and bananas fortified with high levels of essential nutrients. These staple foods dominate diets in many developing countries but lack key vitamins, minerals, and other nutrients. (Gates Foundation, 2005, p13)

'Grand Challenge No.9' (GC9) aimed 'to create a full range of optimal, bioavailable nutrients in a plant'. While distinct from HarvestPlus, Bouis sees GC9 as a HarvestPlus 'spin off'.[31] In contrast to HarvestPlus, however, which at least in its first phase concentrated on raising micronutrient levels through conventional plant-breeding methods, the four successful consortia bidding for the GC9 grants had all proposed transgenic research projects. Furthermore, while HarvestPlus aimed to enrich a range of crops with single micronutrients, the GC9 grantees had the more ambitious goal of stacking multiple nutrients into a single plant.

Almost as soon as HarvestPlus began its operations, therefore, it became clear that it was just one player in a fast evolving landscape of international research. Or, as one BMGF representative commented, HarvestPlus was 'not the only game in town'.[32]

Establishing HarvestPlus

For the HarvestPlus programme team, 2003 was 'the year of getting organized'.[33] This was according to a programme structure that had been set out in the proposal, as follows. Notably, activities would be 'organized by crop', reflecting the core organizing principle of the CGIAR network of research centres:

> A governance and oversight mechanism, led by CIAT and IFPRI, is intended to facilitate these more complex collaborative arrangements ... An external, inter-disciplinary Program Advisory Committee (PAC) of experts from developing and developed countries is being formed to recommend strategic research priorities, oversee project progress, and implement a transparent competitive grants process ... A program leader, a breeding and biotechnology coordinator, and a nutrition coordinator, comprising a three-person Program Management Team (PMT), will coordinate the overall project ... Program activities will be organized by crop, under crop team leaders responsible for coordination. Regional and cross-crop coordination will be facilitated by the PMT and the relevant crop team leaders. (CIAT and IFPRI, 2002, piv)

A series of programme planning meetings followed. In October of that year, the programme name was changed from the Biofortification Challenge Program to 'HarvestPlus', a name considered 'more appealing to the general public'.[34] In a presentation to the CGIAR AGM, the name change was explained as follows:

We took the decision to change the name of the Biofortification Challenge Program to HarvestPlus as a way to reach out more effectively to the public. We felt that this was important in terms of (i) sustaining donor support for a 10-year program, (ii) defending/explaining controversial activities related to development of transgenic crops, and (iii) meeting one of the goals of the Challenge Programs to raise the public profile of the Future Harvest Centers. Not everyone agreed with the decision; several scientists were reticent to use such an 'imprecise' title. However, the decision-making process was highly participatory, the decision approved by a large majority, and accepted and behind us.[35]

During this time, the HarvestPlus network was extending – through competitive bidding processes – to new partner institutions that had not been a part of its precursor projects. Concerns were raised that this process may be straining, rather than strengthening, interdisciplinary relations:

Interdisciplinary exchange/communication is crucial for the success of HarvestPlus. Such interactions become increasingly productive as experience is gained, that is over time and at a series of meetings. HarvestPlus has some advantage that experience was gained by a subset of the collaborating institutions in precursor projects, but many new non-CGIAR collaborators have participated in the planning meetings in 2003 ... Understanding across disciplines is hindered by technical language which either is not commonly understood, or has different connotations to different disciplines ... This all takes time and the give and take of interacting on repeated occasions ... The optimal situation in terms of team building is one in which the partner institutions are all known at the start of the planning process. Competitive bidding can hinder this process of team-building.[36]

By 2004, a range of collaborative arrangements and programme activities were initiated or under way. Notably, HarvestPlus and a new strategic partner, the International Atomic Energy Authority (IAEA), jointly issued bids for bioavailability studies using stable isotopes, with grants awarded to three US land-grant universities: Iowa State, Wisconsin-Madison and University of California (UC) Davis. In addition, a 'reaching end-user' component was added to the programme, with additional funds sought (and later secured) from the BMGF for this purpose.[37] For the HarvestPlus project management team 'the time came sooner than we thought. [At a meeting in] June 2003 the end-user people said "you have to start now"':[38]

In the original proposal, it was an important oversight to postpone collaborative interaction with institutions/implementers related to 'Reaching and Engaging End-users' (that is, moving the biofortified varieties from the research station to consumers) until after nutritionally-improved varieties had been tested and proven for their potential to improve micronutrient status and [were] ready for distribution. Additional funding is needed (and is being sought) for these activities, especially in view of 'fast-track' opportunities in particular for disseminating orange-flesh sweet potatoes, and to some extent for high-iron beans and high-zinc wheat as well.[39]

However, while orange sweet potato was ready to be 'fast-tracked' (with a product inherited from predecessor projects such as Vitamin A for Africa – VITAA), in the event, high-iron bean and high-zinc wheat were not. For example, varieties identified as high-zinc in Peru appeared to lose this trait once transferred to Africa, raising questions about GxE[40] effects, so it was 'back to the drawing board'.[41] At this stage, therefore, HarvestPlus reaching end-user activities went ahead for just one crop, orange sweet potato.[42]

During the course of this 'year of getting organized', therefore, a number of tensions built into in the HarvestPlus Challenge Program design began to surface. The extension of the network to new collaborators, often through processes of competitive bidding, jarred with implicit notions of partnership that had governed relations within the smaller networks that had nurtured a vision of 'scaling up' biofortification research for many years. At the same time, previously accepted – if not wholly resolved – interdisciplinary relations that had evolved within the more informal environment that characterized these networks were unsettled by the new arrangements.

Compounding these strained relations was a new focus on 'impact', now explicit in the directives of the Science Council and the BMGF, at a point when research was at too early a stage to demonstrate it. In light of this, the expectations attached to the 'fast-tracking' of certain projects, and the implications of their mixed success, became increasingly problematic. It is in light of these new tensions and pressures that we return to the Philippines and pick up the threads of the iron rice (IR68144/MS13) and Golden Rice stories, outlined in Chapters 2 and 3, which, though separate until now, were beginning to converge.

HarvestPlus Comes to IRRI

The first HarvestPlus rice crop meeting was held at IRRI in October 2003,[43] convened by Swapan Datta, who now combined his new role as HarvestPlus rice crop leader with his existing responsibilities for Golden Rice adaptive research at IRRI. At this point the iron rice bioefficacy study, using IR68144, supervised by Glenn Gregorio (IRRI) and Angelita Del Mundo (UPLB), was in its final stages and initial findings regarding biological impact were positive (see Chapter 2). This emerging evidence of 'proof of concept' was well timed for the new HarvestPlus programme and the high-iron rice project 'was fast tracked to get data out, to say, there, it's a viable thing'.[44] This was in line with a 'dual approach' discussed earlier, in which the 'early development of "fast-track" varieties that would convincingly demonstrate the validity of the biofortification strategy' was combined with 'a more lengthy parallel development of varieties combining the best nutritional and agronomic traits in each crop' (HarvestPlus, 2004c, p5).

By 2005, however, a number of changes were taking place. After the initial boost delivered by the iron rice study, IR68144, now released by the Philippines National Varietal Improvement Board as MS13,[45] was proving something of a liability, as more cautious, qualified assessments emerged (Padolina et al, 2003). As discussed in Chapter 2, just as these 'black boxes' were reopening, the research 'family' that had, until this point, been steering the project was beginning to disperse. This left a vacuum, as those remaining found themselves caught up in a numbers game, with different stakeholders in the iron rice project – old and new – each advancing their own understanding of the project findings and their implications.

In the same year, as discussed in Chapter 3, IRRI took the controversial step of recruiting Gerard Barry[46] as Golden Rice network coordinator, head of intellectual property and HarvestPlus rice crop leader. These events again shone a spotlight on a project that had come to represent, for IRRI, the promise and dilemmas of rice biofortification: Golden Rice. According to critics, Barry's appointment signified both a point of departure for IRRI as a public research institute and a thread of continuity in terms of Barry's (and, by extension, the biotechnology sector's) involvement with the Golden Rice project. While at Monsanto, he had played a key role in negotiating free licences for the company's IP claims in relation to Golden Rice:

> To critics that degree of control over the introduction of Golden Rice into Asia is merely a continuation of Gerard Barry's Monsanto trajectory – a trajectory that can be traced from the time the corporation realized the incredible PR potential of Golden Rice. According to Ingo Potrykus, 'only few days after the cover of "Golden Rice" had appeared on TIME Magazine, I had a phone call from Monsanto offering free licenses for the company's IPR [intellectual property rights] involved. A really amazing quick reaction of the PR department to make best use of this opportunity.' Barry played a key role in the subsequent negotiations – negotiations which in the end drew in a further five biotechnology companies keen to follow Monsanto's example.[47]

As discussed, Barry's arrival coincided with the departure of Datta from IRRI and from the Golden Rice project[48] – a project which Datta had shepherded from its origins in the ETH laboratory to the IRRI campus in 2001 – and his replacement as the lead transgenics scientist by Philippe Hervé, joining IRRI from CropDesign,[49] a leading agricultural biotechnology company.[50] Hervé oversaw the refurbishment of IRRI's transgenic laboratory at a cost of US$1.5 million.[51] He was joined in 2007 by Susanna Poletti, a postdoctoral fellow arriving from ETH, Zurich, specializing in transgenic approaches to iron biofortification.[52]

With the departure of Gregorio in January 2006 to take up a new position at WARDA,[53] the high-iron rice 'family' had fragmented to the point that it was no longer the voice of high-iron rice at IRRI. Responsibilities for high-iron rice breeding and Golden Rice adaptive research were transferred to Parminder Virk who, prior to joining IRRI in 1999, had spent more than ten years as a research scientist at Birmingham University in the United Kingdom.[54] By 2006, therefore, a new 'international' team had replaced the essentially Filipino research network at the heart of the earlier rice biofortification efforts. In 2006, IRRI published its Strategic Plan for 2007–15. For the first time, a strategic goal (and associated programme) 'to improve the nutrition and health of poor rice consumers and rice farmers' was included, incorporating an objective on biofortification (IRRI, 2006, pp32–3). Barry was appointed to lead this programme.[55]

These staff changes represented an almost wholesale shift from a Filipino research 'family', oriented towards field-based research, to a more impersonal, international network, emphasizing laboratory-based science. While located within the same institutional setting of IRRI, these respective networks lived in different epistemic worlds. In this case, of all the loose ends left open by the departing iron rice family, it was the issue of iron content – its precise measurement and optimization – that would exercise the minds of the members of this new network. In light of these epistemic shifts, we now return to an issue central to HarvestPlus, the creation of interdisciplinary relationships. The following section focuses on two interdisciplinary struggles that took place around this time.

Interdisciplinary Encounters

As already discussed, HarvestPlus is both a formal programme and 'an alliance committed to an idea, a common goal'.[56] One commentator has acknowledged that, 'Howdy [Bouis] did an amazing job, creating partnerships without offending anyone ... He had a lot of knowledge and a vision but didn't impose it, instead worked with [people] created *partnerships* ... created buy-in'.[57] While key to building support for and credibility of HarvestPlus as an interdisciplinary initiative, such alliance building would later generate challenges at moments when these evolving partnerships were called upon to reach 'consensus' in uncertain, unfamiliar or contested areas.

It is at these moments that the tensions inherent in the assumed 'IPG nature' of biofortification research surface, rendering the actor-networks and attempts at black-boxing visible. The following sections highlight two such moments. The first of these focuses on negotiations that took place at the 'centre' of HarvestPlus over programme-wide breeding targets. The second brings the implications of this first encounter back to iron rice research at IRRI, which was now in the hands of the new international network.

Negotiating breeding targets

A critical interface within HarvestPlus existed between agriculture and nutrition, specifically between plant breeders and nutritionists. These tensions surfaced during the negotiation of programme-wide breeding targets: the target levels of micronutrient content set for each HarvestPlus crop that would direct plant breeders in the respective breeding centres. With HarvestPlus framed as a public health intervention, the principle had been to set breeding targets at a level that would achieve a significant biological impact on the health of targeted populations. However, it soon became clear that these targets would be difficult to achieve within the time frame of the programme's first phase. A debate ensued as to whether to put in place intermediate targets, at 50 per cent of the final target, as a first milestone. As one HarvestPlus member recalled: 'Nutritionists set the targets ... Then plant breeders breed to those targets. [But then the question came up] what if we cannot breed to those targets? ... the question of incremental targets ... HarvestPlus has decided to allow that. [But it's an] ongoing question: to what extent can it be incremental?'[58]

This description of an ongoing, open decision-making process differs from the interpretation of a nutritional specialist who had been involved in the project. In contrast, the following account highlights the critical role of power relations between the disciplines, each with their respective institutional underpinnings (or lack of them), in shaping interdisciplinary outcomes. In this case, though HarvestPlus is ostensibly a public health initiative, driven by nutritional targets, these early interactions reveal how these ideas-in-principle were taken up and then transformed through the organizing principles of an agricultural research system constituted from single crop breeding centres:

> Whenever nutrition throws up a challenge, they move the goalposts! [In discussions about levels needed for biological impact] plant breeders realized how difficult ... then the debate about intermediate targets came up. [But] you either have the level for biological impact or you don't. If intermediate targets are allowed ... two risks of running on intermediate varieties ... [first] people conclude nutrition doesn't work ... or [second] they conclude nutrition isn't important ... [The idea that you] just need to breed up shows a huge weakness ... ignores the social science side ... what makes people tick?[59]

Nutritionists had warned of the longer-term consequences of allowing varieties bred to the lower targets to be released and certified as 'biofortified' crops.[60] The case of IR68144, discussed in Chapter 2, demonstrated how expectations could be built up around a – still experimental – rice variety, despite its questionable 'high-iron' status. Nevertheless, plant breeding is a long process – developing a new variety takes several years – so the decision to allow intermediate targets enabled plant breeders to proceed, apparently on their own terms. At this point, the crucial concession was made to the customary practice of 'breeding up', through a series of incremental, achievable steps, towards desired targets – in this case based on nutrient levels.

The decision to 'shift the goalposts' and incorporate intermediate breeding targets into programme plans (see Pfeiffer, 2006) was a turning point in the unfolding of HarvestPlus as an experiment in interdisciplinary collaboration. These events followed a similar pattern as in the formative stages of the CGIAR micronutrients initiative, discussed in Chapter 2, in which an inter-disciplinary research agenda 'tailoring the plant to fit the soil', in which new questions about GxE interactions were at the forefront, was reduced to the straightforward enhancement of required micronutrient levels 'in the seed' through genetic means.

Through a similar set of dynamics, a complex set of questions about plant–human interactions, though central to the issue of impact at the bodily level, was removed from the table by this concession to plant breeders' normal ways of working. The significance of this concession, while played down by IFPRI-based HarvestPlus staff, was not lost on nutritionists. As a boundary negotiation (Gieryn, 1999), the settlement over intermediate targets served to de-link plant breeders' efforts from programme priorities regarding the impact on human nutrition and health. At the same time, these developments ensured that the programme remained dominated by a plant-breeding agenda which, as discussed in Chapter 2, continues to pervade the CGIAR system.

Iron rice revisited

As outlined in Chapter 2, iron rice research undertaken under the ADB-funded programme had been acknowledged for its 'proof of concept' value, and, in particular, its provision of an 'iron bioavailability number for the Filipino diet', a tentative claim that was subsequently reinterpreted as so definitive that 'no more money [would] be spent on it'.[61] Beyond this, IRRI officially had distanced itself from IR68144 as a product and, in particular, from the promotion of MS13, and moved on to screening other varieties for higher levels that would (at least) meet the intermediate targets, now set centrally through HarvestPlus. In continuing the search for suitable breeding parents, the IRRI scientists now working on the project established a boundary between the former breeding work, which screened rice grains in their unmilled form, and their own approach, which was to screen after polishing: 'Most of the work Glenn [Gregorio] did was on brown rice. Now we've moved on to polished rice.'[62]

As mentioned earlier, this focus on the isolation and optimization of grain iron content appealed to a newly established international network that had formed around the iron rice research at IRRI, following the dispersal of the iron rice 'family'. New machines were purchased and refitted to facilitate higher levels of precision.[63] This new focus precluded the exploration of other analytical loose ends that the new iron rice team had inherited from earlier research. The most obvious of these related directly to the boundary-in-the-making between brown and white rice.

One of the lessons from the preparatory studies undertaken by Gregorio and his colleagues prior to the bioefficacy trial was the critical role of post-harvest practices 'in enabling the grain to retain its Fe [iron] content' (Gregorio et al, 2003). In light of this, research into post-harvest practices and, in particular, the merits of alternative milling strategies offered an alternative, or complementary, avenue of enquiry. Given that much of the iron content of the rice grain resides in those parts that are removed by the milling process, some IRRI scientists suggested this as a surer route to achieving the iron levels required.[64] In the event, establishing a boundary between former work on brown rice and the HarvestPlus research on polished rice effectively ruled out this option, at least using HarvestPlus funds:[65]

> We only measure it on polished rice, [we're] not interested in brown or under-milled ... when people want quality rice they want it milled to extreme whiteness, it has to be in that form ... Well there's some interest in it, but we're not going to spend resources on that. Brown rice is more expensive. If people are eating brown rice already they don't have many problems ... we are targeting populations who want ... polished rice. To try to convert people to under-milled rice, when millers want a very tight process ... people pay for high quality whiteness, so trying to figure out intermediate milling, like a default or accidental situation is not something we're interested in spending energy on.[66]

This choice of focus, however, shone an uncomfortable spotlight back on to the problem of iron content. Since previous studies had confirmed that much of the grain iron content was lost in the milling process, IRRI scientists began to express doubts as to whether HarvestPlus breeding targets, even the intermediate targets, would be achievable through conventional breeding methods.[67] Following this reasoning, a dual strategy was chosen: to begin exploratory research on transgenic approaches in parallel with ongoing conventional breeding work.[68] In this case, the programme-wide breeding targets would act as the impartial arbitrator between the merits of the two research pathways. Thus, breeding targets that had been the subject of such heated debate among actors in at the 'centre' of HarvestPlus were now understood as the product of interdisciplinary 'consensus', the last word on nutrient levels required to achieve the necessary impact.[69]

While conducted in parallel, the two iron rice research pathways – employing plant breeding and transgenic approaches – were being conducted under very different sets of pressures and constraints. While the conventional breeding work was behind schedule (having been diverted for some time by the apparent success of IR68144) the transgenic work did not have to produce results until well into its second phase. Furthermore, IRRI scientists have had the opportunity to learn lessons from the experience with Golden Rice, 'controlling their own starting material' to ensure it is 'IP free' and 'marker-free'.[70] Moreover, scientists had some reason for optimism regarding the potential bioavailability of iron in transgenic high-iron rice in light of emerging findings demonstrating that bioavailability of iron in ferritin[71] – the chosen source material – is comparable with that of iron in ferrous sulphate, a common fortificant (Davila-Hicks et al, 2004).

The transgenic iron rice research was taking place in the newly refurbished transgenics laboratory. Everything about the design, layout and aesthetics of this new laboratory suggests a clean break with the past and an attempt to establish a new working culture – a culture within a culture. As a former employee remarked:

> I'm surprised that it's blue. The colour of IRRI is varnish ... varnished wood and white paint, not blue! [The new lab] is renovated for genomics ... from tissue culture and genetic engineering to genomics and DNA. Formerly [we were] just producing transgenic [materials], we didn't know about the expression, whether it's stable. [In the old lab different activities were] all in one room. In the new lab there's one room for each step – DNA, tissue culture [to prevent] contamination. [In the old lab] the library was at the back ... we used to have parties there!'[72]

Debates about whether iron rice research at IRRI would have to 'go transgenic', however, removed from the frame a set of enduring questions about biofortification research; questions that began with its early conceptualization in terms of a food systems paradigm. As discussed in Chapter 2, the early definition of biofortification as a matter of 'tailoring the plant to fit the soil' (Bouis, 1996b, p5; Bouis, 1995b, p18) implicitly acknowledged the role of GxE variation as a dynamic to be exploited, rather than simply observed, as in the case in much conventional plant breeding practice (Simmonds, 1991).

However, a statement by one of the scientists in the international iron rice network that iron content could be treated as any other quantitative trait, and thus GxE would be taken care of through normal multi-locational testing procedures, once optimal cultivars had been identified, indicated that the continuation of plant breeding for enhanced iron levels was following business as usual in formal plant breeding research. This was despite outstanding questions about the role of environmental variation in determining grain iron content, to the extent that, as discussed in Chapter 2, experienced scientists from other crop science disciplines had gone as far as to question whether the genetic half of the equation was at all significant in determining iron content.[73] Others have suggested that, at the very least, research into the environmental and cultural factors optimizing expression of iron might assist in the initial identification of suitable breeding parents.[74]

Nevertheless, discussions at IRRI bypassed these more nuanced debates and, instead, polarized around the 'transgenic or not' question. In this case, the ultimate outcome was seen as resting on whether plant breeders would be able to 'weave their magic' (Gregorio, 2006, p80) and generate a conventionally bred variety that would reach the required target in a race against the clock. If they were unsuccessful, it followed that transgenic iron rice would inevitably emerge as the single pathway. However, as the next section reveals, this was not simply a choice between the (objective) merits of two opposing technologies. To understand what was at stake in these debates, it is necessary to return to the issues of evolving institutional cultures and research partnerships in practice, as providing the context within which members of international research networks, such as those forming around iron rice, make such choices.

Brokers or Gatekeepers? Organizational Tensions and 'Global Science'

Biofortification initiatives evolving during the 1990s had done so in a context of 'longstanding principles of decentralization and centre autonomy' within the CGIAR – principles that in the 2000s had been identified as barriers to building strategic partnerships, especially with the private sector (IFPRI, 2005, p6). Initiatives around orange sweet potato (by the Centro Internacional de la Papa [International Potato Center] – CIP) and high-iron rice (by IRRI), for example, were clearly located in single centres, working with local partners, towards goals shaped by national or regional contexts and priorities. Meanwhile, Bouis had provided a level of coordination and support (within a modest budget). The success of HarvestPlus in gaining approval as a Challenge Program was therefore a point of departure – away from its earlier decentralized practices (and albeit qualified successes) – towards an as yet unrealized ideal, conceptualized as the CGIAR's 'comparative paradigm'. More than this, the role of Challenge Programs such as HarvestPlus would be to provide 'proof of concept' for this paradigm (Science Council and CGIAR Secretariat, 2004, p7).

A comparison between the ADB-funded high-iron rice research at IRRI and evolving rice biofortification research under HarvestPlus is instructive here. As Chapter 2 highlighted, earlier iron rice research was characterized by a research 'family', grounded in a local context and focused on field level results and impacts. In this case, an interdisciplinary team grappled with a series of uncertainties and surprises, responding to these in ways that reflected the context and location of their research subjects and intended beneficiaries. In contrast, biofortification research under HarvestPlus was directed by targets set in a distant location for an anticipated impact on imagined populations. In this case both research avenues – transgenic research and conventional plant breeding – were being pursued with reference to these remote targets rather than the 'field' on their doorstep.

At the same time, this centralizing tendency written into HarvestPlus was reshaping relations between IRRI and its partners. Scientists within IRRI and participating NARS contrasted the lateral relations based on mutual trust and camaraderie that had characterized the ADB-funded project[75] with the sudden relegation of NARS scientists, 'when HarvestPlus came in',[76] to a role in which they conducted field trials and duly submitted their results, but were not party to subsequent discussions. As one NARS scientist explained:

> [The set up] was instituted from IRRI [now we] have to identify donors based on polished rice. Now, for almost a year [we] have been conducting replicated trials for varieties with high iron and zinc. IRRI identified the varieties and composed the nurseries. [After the trials we] submitted grain samples to IRRI for testing. But unfortunately, until now ... no results of the grain analysis ... so ... we can't do the hybridization yet ... don't know from among these materials [which] has high iron ... waiting for more than one year.[77]

An observer from the nutrition community made a similar observation, linking these emerging dynamics to cherished notions about CGIAR centres as 'centres of excellence': 'The proportion of HarvestPlus funding going to national institutes is *miniscule*. HarvestPlus is not developing the NARS to do the work. That's the challenge of interdisciplinary work ... [the] CG mentality [is] *"we are the centres of excellence"'*.[78]

As discussed in Chapter 1, in Challenge Programs such as HarvestPlus, the CGIAR sought to bridge disciplines from the 'classic cluster' of crop sciences (Anderson et al, 1991, p74), particularly plant breeding, with human science disciplines that lay outside the confines of its expertise. In this case, notions of the CGIAR centre as a 'centre of excellence', while arguably already outdated (Chataway et al, 2007), had become more problematic. However, while official HarvestPlus discourse highlights interdisciplinary collaboration, with CGIAR centres acting as 'brokers' rather than experts, the experiences recorded above suggest a reluctance to relinquish this cherished position, so the role of broker is transformed into that of gatekeeper.[79] As discussions in Chapter 3 have revealed, a precedent for this role-shift already existed in the shape of the Golden Rice project (by now, as discussed in Chapter 3 and in the later sections of this chapter, in the process of extending to the ProVitaMinRice Consortium), in which a highly hierarchical set of 'partnership' arrangements had become normalized over time.

These tensions have implications for the practice of science, both within CGIAR centres such as IRRI and its national partners. Sumberg (2005) has drawn attention to the way in which incentives and pressures channelling NARS efforts and limited resources towards international collaborative projects are diverting these national institutions away from research that might respond to local agro-ecological conditions. These dynamics are intensified in the Philippine context, given the late entry of PhilRice into the community of NARS in the region, in the context of accumulated disappointments and misunderstandings over the expected contribution of IRRI to Philippine agriculture (Perlas and Vellvé, 1997). In view of this legacy, and the current reality in which NARS are pulled into international 'collaborations' as a way of attracting much needed funds, these dynamics within HarvestPlus reinforced these asymmetric relations and provided yet another reminder that PhilRice remained, in effect, 'IRRI's younger brother'.[80]

This shift has been accompanied by transformation to a more 'top down' approach to the conduct of science, as the iron rice project changed hands from a grounded, located iron rice 'family' to the laboratory-based realities of the international network, that has reshaped iron rice research under HarvestPlus. For IRRI's new iron rice network, context-responsive, 'bottom up' research has no place in a programme that has retained aspects of the traditional 'centre of excellence' role, while deferring to a global framework and set of targets whose empirical relationship with projected improvements in human health and welfare is no longer questioned. As one IRRI scientist asserted: 'Bottom up ... doesn't work ... [It results in] messy measurement ... [which you] cannot replicate. In biotech, we are used to a lot of controls. For example ... iron content ... at IRRI ... everyone measures iron in a different way. This is the difficulty with bottom up. I can't understand why such a simple thing ... so important ... cannot be standardized'.[81]

This privileging of the universalizable over the site-specific (Biggs and Clay, 1981) – from the formalization of central breeding targets to the treatment of GxE uncertainties within iron rice research at IRRI – has, of course, been an enduring theme in the CGIAR trajectory, as discussed at length in Chapter 1. Of particular note is the way in which IRRI had provided the platform for a new actor-network to selectively appropriate and reframe an earlier initiative, of which a locally based research 'family' had taken such ownership, to assume membership of an evolving global science community. Implications of these shifts for the practice and quality of science – in particular the intensification of black-boxing of uncertainties that has proliferated under such circumstances – have been explored in this section. The next section discusses emerging policy and impact questions.

Constructing Demand, Predicting Impact

The status of HarvestPlus as a pilot Challenge Program incurred built-in expectations that it would prove to be an exemplar of the international public goods model. These expectations are ambitious, given the multiple uncertainties inherent in biofortification research. While building on prior experiences inside and outside the CGIAR system (Low et al, 2001; Corpuz-Arocena et al, 2004; Johnson-Welch et al, 2005; Haas et al, 2005), on closer inspection these earlier projects raised as many questions as they answered. This was not surprising given the complex ecological, socio-economic and cultural dynamics that mediate local nutrition outcomes. In one of the papers that launched the CGIAR micronutrients initiative that preceded HarvestPlus, Calloway cautioned that: 'the target population is not homogeneous so one remedy is unlikely to serve all. To intervene efficiently and effectively requires knowing fairly precisely what a population lacks and why. That knowledge is no less necessary for selecting crop-modification strategies than for formulating policy' (Calloway, 1995, p21).

The definition of 'Reaching and Engaging End-users' in HarvestPlus literature as a matter of 'moving the biofortified varieties from the research station to consumers' (HarvestPlus, 2005a) indicates that Calloway's concerns have gone unheeded. Instead, the notion that consumers are to be faced with a straightforward choice of whether to 'switch' from non-biofortified to biofortified crops – for part or all of their total consumption – has gained currency within HarvestPlus: 'Two basic assumptions are possible ... (i) a certain percentage of consumers switch completely to the biofortified crop, while the remainder consume only the unfortified crop, and (ii) all consumers replace a certain percentage of their consumption with the biofortified crop and continue to consume some unfortified crop in parallel' (Stein et al, 2005, p20).

The way in which the 'end-user' question has been framed in terms of consumer choice;[82] whether, and to what degree, to consume fortified and/or unfortified rice is reminiscent of early assumptions underpinning the roll-out of Golden Rice, as discussed in Chapter 3. These assumptions endure, despite the characteristically heterogeneous and segmented nature of rice markets in Southeast and South Asia, in which a multiplicity of varieties, milled to varying degrees, coexist in a 'marketplace' populated by numerous small-scale rice farmers, millers and vendors.[83] Furthermore, as discussed in earlier chapters, instrumental 'consumer choice' models fail to take account of the myriad ways in which the cultivation, harvesting, preparation and consumption of rice shape and reflect social and cultural life (for example, see Hornedo, 2004, p5).

The decentralized nature of all aspects of rice markets therefore compounds the essentially local character of nutritional outcomes, which are invariably mediated by local biological and ecological conditions in addition to the socio-economic and cultural factors that shape the production and consumption of food. Furthermore, while ostensibly 'food-based',[84] biofortification is nevertheless 'a health intervention, using food as an intervention to administer extra nutrients'. In this case, dissemination of biofortified varieties calls for a specificity and rigour in targeting and needs assessment beyond that normally required from the agricultural research community: 'As a health intervention, there are implications ... first you must first identify the need ... target group, and second [it] needs regulation ... in which the Ministry of Health has something to say. [When there is a clearly defined need] then you can justify your risk'.[85]

Yet these realities are at odds with the imperative of HarvestPlus to produce IPGs that will generate demonstrable 'impact'. Instead, HarvestPlus documents refer to constructed 'populations' such as the 'nutritionally disadvantaged' (CIAT and IFPRI, 2002, p5). Groups such as the poor, particularly those living in rural areas of developing countries (CIAT and IFPRI, 2002, pp4–14), and women and children in particular, are assumed as beneficiaries (Stein et al, 2005). In this case, HarvestPlus is envisaged as extending the 'choice' available to the 'poor' and 'nutritionally disadvantaged' as individuals in the marketplace, in an economistic vision of technology–society relations.

HarvestPlus has been presented as inherently 'pro-poor', by targeting staple crops, since the diets of poor people contain proportionally more staple foods, and by its strategy of plant breeding, which can reach further into remote rural areas than a strategy based on post-harvest food processing (CIAT and IFPRI, 2002). In this case, it is assumed that incorporation of biofortification into the national varietal release mechanisms in developing countries will automatically produce pro-poor outcomes.[86] Similarly, the increased micronutrient requirements of certain categories of women and children are translated into beneficiary status (Stein et al, 2005). However, it is well established that the special requirements of these groups call for concentrations of micronutrients, such as those available in pharmaceutical supplements and micronutrient-dense complementary foods, and are unlikely to be offered by biofortified staples.[87]

This construction of dislocated, generalized populations complements the construction of the end-user as rational consumer, to provide the baseline for the elaboration of an econometric approach to predict impact, *ex ante*. This approach is an adaptation of a framework originally developed by Roukayatou Zimmermann at the Zentrum für Entwicklungsforschung (ZEF – Center for Development Research), University of Bonn, and Matin Qaim at the University of Hohenheim, to present 'the potential health benefits of Golden Rice' (Zimmermann and Qaim, 2004). Qaim, one of the more recent recruits to the Golden Rice Humanitarian Board,[88] has co-authored a number of papers extending an adapted version of this framework to inform the approach to 'impact and policy' issues within HarvestPlus (Stein et al, 2005; Stein et al, 2006).

The sophistication of this economic analysis, however, belies its reliance on simple causal pathways linking anticipated shifts in 'consumer choice' with expected health outcomes; then extrapolating economic impact, using the 'standard' epidemiological unit of the disability-adjusted life year (DALY, see Chapter 2):

> To measure the economic impact of biofortified staple crops on public health, both the number of DALYs lost under a hypothetical scenario, in which people consume biofortified crops, needs to be calculated. In addition to the information needed to calculate DALYs under the *status quo*, developing a hypothetical scenario where people consume biofortified crops requires further information; specifically, the contribution of biofortified crops to a reduction in micronutrient malnutrition, and hence to an improvement in public health, needs to be specified. (Stein et al, 2005, p8)

In this case a rationale constructed to justify the roll-out of Golden Rice as a product has been extended to inform impact analysis for a range of biofortified varieties planned under HarvestPlus. While couched in the language of a public health and pro-poor development, HarvestPlus continues this tradition of 'reaching end-users' with predetermined products in such a way as to reinforce assumptions that international biofortification efforts have potential to generate impacts from outputs of an 'IPG nature'. This approach to conceptualizing impact combines the narrowing effects of two disciplinary lenses, those of crop science and agricultural economics. Both have contributed to a blind spot within the international biofortification enterprise in general, and HarvestPlus in particular, around the ways in which these efforts are likely to have an impact (or not) on the health and welfare of actual human subjects.

However, as discussed in Chapter 1, this approach to 'impact' reflects global trends in which dominant ways of thinking about nutrition, health and development privilege centralized, global goal-setting, standardized epidemiological analysis and an overriding concern with cost-effectiveness and the removal of constraints to individual productivity. In this case, this approach to the question of impact is part of a global biopolitics in which the universalizing assumptions of agricultural science, with its overarching plant genetics frame from which the international public goods model is derived, and international nutrition, with its predisposition towards a 'fixed genetic potential' approach (Pacey and Payne, 1981), combine with the individualizing effect of neoclassical economics to detach constructed populations from any sense of geographic, socio-cultural or political location.

Impact and 'Spin-Offs'

As HarvestPlus neared the end of the first phase, it was becoming clear that programme outputs would be limited. Delays had been experienced, not only in rice, but also in the wheat and bean adaptive research.[89] While the orange sweet potato component was progressing well, enabling the programme team to experiment with 'reaching end-user' strategies, the project infrastructure and outputs had largely been inherited from the earlier USAID-funded VITAA (Vitamin A for Africa) project (Low et al, 2001). By early 2006, donors were expressing concerns about the time it was taking biofortification to 'come to scale': 'We've been hearing about biofortification for a long time but it still hasn't come to scale ... [they say it] will be another five to ten years ... What's holding it back?'; 'Orange sweet potato is the only success story, and that's questionable.'[90]

Given these concerns, how would donors assess the impact of HarvestPlus at the end of its first phase? Would their support continue into the second phase? At this point, the focus of attention for the HarvestPlus project management team shifted from a fragile 'core' to the prospect of generating 'spin-offs' as a more genuine measure of impact. In this case, attention focused on emerging national biofortification programmes initiated in India, Brazil and China (Bouis, 2006).[91]

Taking the example of the HarvestPlus-China programme: this was conceived in 2004 'after the active communications among Dr Howdy Bouis,

Professor Yunliu Fan and Professor Xingen Lei'. Fan's enrolment, as a 'respected academician of the Chinese Academy of Engineering Sciences' has proved crucial to gaining the support of Chinese scientists to a programme that had yet to secure substantial support from the Chinese Government, as was that of Xingen Lei of the nutrition faculty at Cornell University who, as a respected Chinese scientist and US-based HarvestPlus collaborator, played an important bridging role. Early organization and progress of the programme are described by HarvestPlus as follows:

> To initiate the program in China, HarvestPlus programme com-
> mitted $350,000 for pilot research for the Chinese scientists. The
> projects were required to be target [sic] on improving Fe [iron] and
> vitamin A bioavailability in rice, corn and wheat, with 1:1 match-
> ing support from home institutes. A total of 16 applications from
> 39 institutes were submitted to the program office on February 1,
> 2005 for initial evaluation. In April 2005, 7 projects were selected
> by the program office and advisory committee for funding. These
> projects will be conducted by multiple institutes, with very specific
> target nutrients (Fe or vitamin A) and population. Several field
> leaders and teams have carried on field investigations and held
> project organization meetings. All laboratories are currently work-
> ing with HarvestPlus program scientists on verifying the nutrient
> analysis in the selected samples.[92]

In September 2007, HarvestPlus-China held its second annual meeting, which I attended. At this meeting, jointly chaired by Fan and Bouis, participating Chinese scientists presented their work. In addition to Bouis, several senior HarvestPlus programme staff, crop leaders (including Gerard Barry as rice crop leader) and collaborators (including Ross Welch and Robin Graham) attended. Notably, Welch and Graham were also members of the HarvestPlus-China Program Advisory Committee, within a programme structure that mirrored that of HarvestPlus.

While Bouis and other external participants were clearly impressed with the scientific output within a relatively short timescale, concern was growing that the Chinese government had so far committed minimal funding. Fan stressed the need for patience, while continuing with a 'do and ask' approach, 'to show outcomes ... get convincing scientific data, then get funds from central government'.[93] For HarvestPlus representatives, however, the imperative was to show that, despite limited progress in generating biofortified varieties to date, the programme had achieved impact through its generation of spin-offs. In this case, the prospect of leveraging funds from the Chinese government was seen as crucial to establishing this argument. HarvestPlus representatives therefore advised the Chinese scientists to take a pragmatic approach towards building the case for funding: 'Start at the end point and work back. Maybe you don't have time to wait for a CACO-2 result.[94] The ownership ... the needs of the project here ... align timelines. [You need to] make certain leaps of faith ... fill the gaps in later'.[95]

At the same time, the extension of the HarvestPlus network to the China programme further consolidated the black-boxing of results that, closer to the core, were still in doubt. IRRI's high-iron rice study, in particular the nutrition study, was recommended as an effective, proven method for strategic use and communication of research to attract funding. In this case the positive aspects – in particular the success of the bioefficacy study[96] conducted in the Philippines in delivering 'proof of concept' and so attracting donor support – were highlighted. In addition, the previously contested 'bioavailability number' drawn from the study was presented as a definitive result: '[You need to] determine the bioavailability number in the Chinese diet. We have it for the Philippine diet. Go back, pick a study that will have a huge effect ... [Use] stable isotopes, you define the diet ... and how you define the communication value'.[97]

The case of HarvestPlus-China suggests an attempt by the HarvestPlus leadership to create a programme largely in its own image. In this case their efforts had limited success, particularly in terms of their chief aim of bolstering the core of HarvestPlus by demonstrating its power to leverage funds from new sources. Instead, the most significant outcome may have been the extension of networks supporting previously contested or qualified findings and approaches, with the result that these were further black-boxed as definite findings and proven methods.

Business as Usual? The ProVitaMinRice Consortium

While attempts to demonstrate the added value of HarvestPlus as a platform for 'spin-off' national programmes generated mixed results, progress of the earliest HarvestPlus 'spin-off' may be more indicative of the dynamics and future direction of biofortification research, in which a close-knit, international network, initially formed around the goal of promoting Golden Rice, was playing an increasingly central role.

Following its decision to fund HarvestPlus in 2003, the BMGF launched its 'Grand Challenges for Global Health' initiative. One of these challenges – Grand Challenge No.9 (GC9) – was to create a full range of optimal, bio-available nutrients in a plant (Gates Foundation, 2005, p13), in other words to produce a multi-nutrient biofortified crop. Four proposals were selected, for research on rice, sorghum, banana and cassava. One of the four grantees was the ProVitaMinRice Consortium (PVMRC), which aims to engineer 'rice for high beta-carotene, vitamin E, protein, iron and enhanced iron and zinc bioavailability'. At the centre of this 'new' consortium were key players from the Golden Rice project, including one of the co-inventors, Peter Beyer, at Freiberg University, as lead scientist. However, in a similar manner to HarvestPlus, the PVMRC has extended to include additional US universities (Michigan State and Baylor), NARS in the Philippines (PhilRice) and Vietnam (Cuu Long Institute), and the Chinese University of Hong Kong. [98]

In addition to vitamin A and iron, which were already priorities within HarvestPlus, this programme introduced two additional nutrients. Lysine – an enduring 'relic' (Bryce et al, 2008, p1) from a nutrition era dominated by the protein paradigm – was again back on the agenda. Vitamin E was introduced, as a nutrient in its own right (though, notably, it is difficult to find any mention of vitamin E deficiency as a public health priority in developing countries) and as an alternative route to solving what has been a persistent problem for the Golden Rice project, the stabilization of beta-carotene: 'Tocopherols and tocotrienols constitute Vitamin E, an essential component of the diet, and have the additional benefit that they help stabilize provitamin A within the food matrix owing to their strong antioxidant properties' (Al-Babili and Beyer, 2005, p571).

In the early stages, consortium members were each pursuing 'proof of concept' research on an individual nutrient or aspect of this complex programme.[99] However, the overall programme was presented as an interdisciplinary, multi-nutrient initiative, since the expectation was that, at a later stage, the genes and constructs produced from this upstream research would be pyramided into the Golden Rice 'finished product'. As one scientist working for the consortium observed: 'It's only Vitamin A that pulls it all together, in relation to rice. Vitamin A is also the common denominator of all staple crops in GC9 ... cassava, banana, sweet potato'.[100]

The centrality of Golden Rice to the design of the PVMRC was reflected in institutional arrangements governing the programme, in which an expanded Golden Rice Humanitarian Board had assumed the role of general oversight, as 'an external advisory board'.[101] As outlined in Chapter 3, this governing body was originally formed for a specific purpose: to facilitate the release of proprietary knowledge and materials according to the terms of a 'humanitarian licence'. However, over time it has grown in size and broadened its mandate, first over the Golden Rice project and now extended to the PVMRC. Meanwhile, the structure and timing of meetings, linking HarvestPlus, PVMRC and the Golden Rice Humanitarian Board, are indicative of continued 'mission creep', together with the emergence of new hierarchies:

> Both HarvestPlus and [CG9] are governed by the Humanitarian Board in terms of research directions. During [PVMR] Consortium meetings and HarvestPlus meetings the Humanitarian Board is there as an R&D board. [In 2005 a series of meetings were organized] back-to-back: The HarvestPlus main meeting ... then the Humanitarian Board meeting ... then the Humanitarian Board meets each [PVMRC member] at a time ... At the last meeting Humanitarian Board came three days after ... We were advised what to present to them ... then we were asked to leave ... so they can discuss.[102]

This convergence of networks challenged distinctions between the different projects – HarvestPlus, Golden Rice and the PVMRC – distinctions that exist on paper but not in practice. In reality, the networks supporting these respective projects are interwoven, with back-to-back meetings an inevitable logistical outcome. Nevertheless, when placed in the broader context of an overextended network, it is the hierarchical character, and the ordering of particular actors within those hierarchies, that raises the most serious concerns. In particular, the omnipresence of the Golden Rice Humanitarian Board as the default governing structure points to the vulnerability of an overextended network to the possibility of fragmentation and capture.

Conclusion

HarvestPlus evolved from the earlier CGIAR micronutrients initiative, a relatively modest programme driven by the commitment of a group of individuals. In the process, it has absorbed a range of regional initiatives within an ambitious global vision, buoyed by a reassertion of the CGIAR's comparative advantage as a generator of international public goods, achievable through a 'return to its roots' in upstream research. Reframed as a Challenge Program, HarvestPlus embodies the tensions inherent in a new way forward identified for the CGIAR, in which the interests and agendas of heterogeneous actors are to be reconciled within 'strategic partnerships' and channelled towards pro-poor ends.

This chapter has followed the evolution of HarvestPlus as a process of network extension and overextension, relying, not on empirical assessments of needs of particular people in particular places, but on the abstraction and aggregation of dislocated populations deemed 'at risk' from malnutrition-related diseases. This is made possible through the overlapping of disciplinary worldviews of a universalizing plant genetics frame and an individualizing, neoclassical economics frame, each of which fails to recognize the location of people, malnourished or not, within socio-economic, cultural and geographic contexts. This 'blind spot' continues to pervade the international biofortification enterprise in ways that those central to it are unable to see. Recognizing the limited achievements of the HarvestPlus initiative in its own stated terms, its coordinators look instead to 'spin-offs' to reflect indications of success back to the core project.

However, underlying these developments is an enduring set of assumptions about the role of CGIAR centres as 'centres of excellence'; assumptions that conflict with the new role of 'broker' that architects of the Challenge Programs envisaged for its research centres. Though nutrition and health are recognized as lying outside the CGIAR's area of expertise, nevertheless, as the interdisciplinary negotiations outlined in this chapter highlight, CGIAR scientists involved in HarvestPlus have been successful in transforming the programme design and goals in such a way as to reassert the traditional 'centre of excellence' status of the CGIAR centres within the programme.

This type of selective interpretation and appropriation, at the level of the breeding centre, of a global agenda intended to transform the CGIAR is, of course, just the kind of dynamics against which Eicher and Rukuni have cautioned (2003, p24). Meanwhile, an overextended network may be in danger of fragmenting, as the core group driving the Golden Rice project plays an increasingly influential role in shaping the boundaries between the various projects that populate the global biofortification landscape. In this case HarvestPlus has a role to play as the public face of biofortification, while 'business as usual' continues.

5
Global Science, Public Goods?
A Synthesis

With the transformation of international biofortification research into a global 'challenge', an attempt was being made to streamline a multiplicity of diverse pathways, at different stages and of varying levels of context-responsiveness into a globalized, high-tech 'fast lane'. This book has, through a series of cases, explored how and why this singular vision for biofortification has survived unscathed, despite a catalogue of practical setbacks and some well-informed critiques coming from inside and outside the CGIAR system.[1] To understand the broader significance of biofortification, in this context, requires an extension of the boundaries of the debate, beyond contesting whether it will 'work', to an analysis of the modes of organization, styles and cultures of science and definitions of impact evolving around it.

These complex dynamics, this book argues, offer a glimpse of the future of international crop research conducted in the name of development and poverty alleviation. International biofortification research, in its current form, exemplifies a particular formula for connecting agriculture, nutrition and health which resonates with now dominant framings of global health and international development, and is leading all three fields towards a 'consensus' around generic, reductive, centralized approaches. This chapter synthesizes lessons from the cases presented in the last three chapters, around the core themes of research organization and partnership, interdisciplinary and uncertainty, and impact and upstream–downstream linkages, and draws out implications for the future of global 'public goods' crop science.

International Research Partnerships: Rhetoric and Reality

The construction of biofortification as a field that inherently crosses disciplinary and sectoral boundaries has led to a profusion of partnerships of various kinds. This is illustrated by the three cases explored in the previous chapters: from a localized research 'family', with some international membership (as in the case of iron rice), to diverse, complex international networks, such as those around HarvestPlus, and particularly as they converge with the overlapping networks around Golden Rice and the ProVitaMinRice Consortium (PVMRC).

Ideas about international research collaboration have been evolving since the establishment of IRRI in the early 1960s, a pioneering move by the Rockefeller and Ford foundations and the government of the Philippines that catalysed the establishment of an international system of agricultural research, the CGIAR. During the 1980s and 1990s, these formative ideas, in which North–South collaboration in genetics-led crop science was identified as an effective means to effect widespread improvements in socio-economic well-being in the South, crystallized in the Rockefeller Foundation-funded International Program on Rice Biotechnology (IPRB). Over a 17-year period, this programme helped to embed a set of relations between institutions in different parts of the world according to a set of relations that counter posed 'cutting edge' expertise in the advanced research institutions in the North with 'need' in countries in the 'rice-dependent' South (O'Toole et al, 2001).

The case of iron rice illustrates the modest beginnings of biofortification research within the CGIAR system. Starting with a series of connections between researchers at the Washington, DC-based IFPRI, PSNL, Cornell University and Waite Institute, Adelaide, the idea of rice biofortification was picked up by a group of researchers in the Philippines, led by IRRI plant breeder Dharmawansa Senadhira and Angelita Del Mundo, a nutritionist at the neighbouring University of Philippines, Los Baños. Both key figures in developing the potential of this new direction in crop research, they oversaw the evolution of a close-knit interdisciplinary network or 'family' of researchers, mobilized by the serendipitous 'discovery' of a high-iron rice variety. What followed was at the same time a landmark nutritional research project, drawing on the expertise and guidance of nutrition scientists in North America and the 'grandfathers' in Washington, DC, Cornell and Adelaide; and a uniquely Filipino blend of science, Catholicism and development. While drawing on international 'experts', however, this was a network grounded in the social realities and priorities of the Philippines, as viewed through the eyes of the research family members.

The iron rice project can be understood as a socio-technical network that evolved over several years, within a particular historical and social context. Nevertheless, this did not prevent its unravelling within a remarkably short space of time. Just as results were emerging, the family was dispersing, their places taken by new arrivals whose professional identity was 'international', rather than Filipino. This changeover coincided with the arrival of HarvestPlus, which was a new packaging of biofortification research as a global project, underwritten by the promise of a new influx of funds. At this point, lessons as to the advisability of a more comprehensive analysis of the place of biofortified rice within national priorities of the day were eschewed by a new set of 'global' actors. Instead, they responded to a new imperative, to isolate and draw from the inevitable messiness of a research project, so deeply enmeshed in its social context, the necessary scientific 'proof' to shift biofortification research up a gear and shore up its status as an exemplar of global, public goods science.

In the process, 'local' actors and institutions have been increasingly sidelined, as new international networks have formed around larger stakes and resources. Discussions with Filipino scientists in different institutions reveal a dissonance between the rhetoric of international partnerships and their own daily experience. They draw comparisons with earlier projects with more modest levels of funding (such as the ADB-funded iron rice project), which they had experienced as more open and transparent, with funding and information flows more regular and predictable, and working relationships more indicative of partnership in practice.[2]

In recent years, the logic of the 'research training-cum-collaboration' model at the heart of the IPRB (O'Toole et al, 2001) has been both reinforced and transformed by a more recent imperative to access the research capacity and resources now concentrated within the transnational private sector. In this case, a language of partnership, collaboration and consensus emphasizes horizontal relations and serves to obscure the hierarchies that have emerged in practice. The functioning of the Golden Rice network is illustrative. Externally, this network appears to mirror established relations between IRRI and NARS in the region. However, IRRI's role as 'technology holder',[3] on behalf of the Humanitarian Board, brings different obligations, generating markedly different practices of knowledge and materials transfer and communication.

In particular, lateral communication between NARS is minimal; instead, vertical relations between IRRI – as the network 'hub' – and individual NARS are emphasized. These dynamics highlight the uneasy coexistence of the elements of 'broker' and 'gatekeeper' within IRRI's somewhat ambiguous role as network hub (Mackintosh et al, 2008). Furthermore, the emergence of parallel research pathways – what could be described as the high road of the laboratories of Syngenta and the low road of the NARS adaptive breeding programmes, with IRRI bridging the two – illustrates the inherently asymmetric structure of these partnerships in practice.

HarvestPlus has been described as 'an alliance around an idea',[4] with CGIAR centres acting as 'brokers' (Rijsberman, 2002) holding together extensive, heterogeneous networks, representing a range of histories, agendas and interests. It therefore has to present itself as the common ground, the rallying point for a diverse network of member organizations, each functioning according to their own 'institutional action frames' (Schön and Rein, 1994, p32). In this case, the role of independent governance structures, established to oversee and guide a programme of this sort, is fraught with unprecedented levels of complexity and ambiguity. Who does drive an amorphous entity such as HarvestPlus? How is it to be held together and at what cost? Developments within HarvestPlus appear to be echoing those within the Golden Rice project, neglecting issues of context-responsiveness in order to maintain complex upstream institutional arrangements in such a way as to retain donor confidence.

Implications of the structure of relations within these emerging partnerships goes beyond concerns about the marginalization of contributions and insights of developing country institutions, first, to the way in which 'development' is conceived and enacted, and, second, to the conduct and practice of scientific enquiry itself. In particular, the assertion of a traditional linear 'pipeline' heuristic for innovation and technology diffusion has served to oversimplify both the technical and policy dimensions of a highly complex endeavour, making it appear far more manageable than it is. At the same time, this linear model has served as a justification for a hierarchical set of relationships that downplay the contributions of developing-country partners. This is despite the positioning of these partners further 'downstream', closer to the end-users, about whose actual needs and preferences in relation to biofortified rice so little is known.

While reassuring to proponents and donors, this mode of thinking implicitly discourages those concerned with scientific enquiry and policy advice from 'looking sideways',[5] and exploring the multiple uncertainties inherent in the biofortification enterprise as a whole. These dynamics highlight an apparent contradiction within these extended partnership arrangements, in which their external appearance, as formations capable of managing high levels of complexity, belies a practice, deeply embedded in institutions such as the CGIAR, of reducing and tailoring complex problems according to its own internal organizing principles, and structuring its relationships with other institutions accordingly.

Towards Interdisciplinary Integration?

Each of the biofortification initiatives explored in this study has attempted to bridge disciplinary boundaries in some way. Earlier chapters have charted events through which such connections were constructed, but in each case subsequently unravelled, exposing a complex of disciplinary language barriers and inter- and intra-institutional asymmetries.

The case of iron rice is instructive of a series of shifts in styles of interdisciplinary science; from an open-ended, exploratory approach in the early days, to the crystallization around the nutritional efficacy study with the 'Sisters of Nutrition', of a more systematic design, aimed at providing 'proof of concept' that would convince national decision-makers and international donors. Just as this proof was in sight, however, the project metamorphosed again into an exemplar of global, cutting-edge science. At this point, the spotlight shifted away from the Philippines, where interest in nutritional rice was eclipsed by national productivity and self-sufficiency objectives activated through a renewed focus on high-yielding hybrid varieties, with a series of realignments of international priorities and resource allocations following the approval of HarvestPlus as a Challenge Program.

A continuous thread running through these transformations, however, has been a series of interdisciplinary encounters within a CGIAR system with a built-in tendency towards reductionism, articulated through the hegemony of plant breeding within a taken-for-granted interdisciplinary hierarchy. Over time, successive interdisciplinary puzzles were 'solved' through the closure of black boxes around outstanding questions about the unexpectedly high variation in agronomic performance of the high-iron rice variety in different agro-ecological conditions, and the factors determining the ultimate nutritional value of those varieties when consumed by particular groups of people as part of their normal diet. While the uncertainties around GxE interaction would be 'taken care of' through conventional multi-locational field trials,[6] nutritional dynamics were reduced to a standard 'bioavailability number for the Filipino diet'.[7] In this way, the key research problem to emerge as the focus of further scientific enquiry was grain iron content and its optimization.

The Golden Rice project has, in contrast, attempted interdisciplinary integration further upstream, with the formation of a boundary-crossing Humanitarian Board to oversee a process of technology transfer via IRRI to a 'Golden Rice network' linking research institutions across South and Southeast Asia. Despite these institutional developments, Golden Rice – in its various, still incomplete forms – has consistently been treated as an almost finished 'product' in a process of being rolled out, according to a classic linear diffusion model that fails to take account of the diversity and particularity of contexts in which such a product – a novel rice variety – might be adapted, grown and consumed.

These technical and policy uncertainties have, however, been framed out of debates that have highlighted the innovative nature of upstream institutional arrangements. These dynamics create challenges for scientists undertaking adaptive and applied research further downstream. Inevitably, outstanding technical uncertainties are pushed downstream, complicating their task in various ways. At the same time, the prior framing of Golden Rice as technically proven restricts the scope for scientists to debate these uncertainties openly.

Earlier discussions, about the type of partnership arrangements that have evolved around Golden Rice, illustrate the emphasis of vertical over horizontal relationships, and the project therefore lacks the kind of open, collegial relations between scientists that had characterized the iron rice 'family'. Notably, the activities of adaptive breeding and nutritional testing, integrated within a single team for the iron rice project, have, within the Golden Rice project, been conducted entirely separately, within different institutions in different countries. Furthermore, the experience of NARS scientists who, after several years' adaptive research, were instructed to destroy their materials and start again,[8] and delays in sharing results of nutritional tests, suggest that the openness required for genuine interdisciplinary exchange is not a feature of the Golden Rice project. Thus, while the composition of the Humanitarian Board suggests a level of interdisciplinary integration within the project, clearly the spaces do not exist for scientists working 'on the ground' to engage in open interdisciplinary exchange.

The HarvestPlus program had the advantage of following the other two projects, and so an opportunity to put lessons learned into practice. However, such lessons have been partial and selectively applied, due to the onset of new pressures. As a pilot Challenge Program, expectations of HarvestPlus were high: that it would provide 'proof of concept' for the Challenge Program model as the way forward for an international research system grappling with ways to secure its financial future, while, at the same time, reasserting its traditional mandate. The Challenge Program model, and associated justifications, illustrated an uneasy compromise between a *raison d'être* of 'international public goods' and an understanding of reality as one in which funds must be leveraged from new sources (including the private sector) if an international public institution such as the CGIAR was to remain viable. In this case, HarvestPlus was the clear success story among the four pilot Challenge Programs, since it was the lone example of a Challenge Program that has resulted in a substantial inflow of new funds (from a new source, the BMGF), rather than a re-channelling of existing funds from traditional sources.[9]

How has the HarvestPlus programme mobilized these substantial resources, given the available lessons from previous projects? The programme language, starting with the 'new paradigm', is indicative of a commitment to interdisciplinarity, partnership and consensus building. In recognition of the limits of single (in particular single crop) research centres, interdisciplinarity is envisaged as the outcome of 'strategic partnerships' between institutions of various kinds. However, analysis of earlier cases, such as Golden Rice and iron rice, reveal as flawed the assumptions of a smooth transition of upstream collaboration into integrative local practice, and highlight the multiple challenges in building interdisciplinary practice on the ground. Furthermore, while the local nature of the iron rice project ensured that any misguided assumptions or interdisciplinary gaps were bound to be exposed – that, in fact, was its strength – in the case of Golden Rice, and now HarvestPlus, the upstream focus provides refuge from such lessons, generating an aura of interdisciplinarity which has yet to be realized in practice.

De-linking Impact and Context

As discussed in Chapter 1, current biofortification initiatives represent just one of a series of attempts to link agriculture, nutrition and health. The significance of biofortification, and the way it is framed within HarvestPlus, is that it fits a formula that, at the present time, is highly influential in guiding large-scale investments in research and development. According to this formula, investments in the type of upstream activity that can be linked, via 'simple causal pathways'[10] and vertical programme designs, to measurable results downstream, are regarded as having the highest potential for 'impact'.

The formula can be traced to the Millennium Development Goals (MDG) framework, now the overarching framework within which development programmes, and research *for* development programmes, are routinely justified; and can also be traced to earlier developments around the conceptualization of global health within a standardized epidemiological framework (Murray and Lopez, 1996). In 2004, this formula received further endorsement from the highly influential 'Copenhagen Consensus', on interventions offering the most cost-effective solutions to 'the world's biggest challenges'.[11]

The reforms within the CGIAR at that time, emphasizing a return to upstream research as a way to increase its impact, were consistent with this formula. HarvestPlus, in particular, addressed the favoured strategy of 'providing micronutrients' through a relatively simple pathway, linking quantifiable increases in levels of micronutrients available from staple crops with projected health and welfare improvements. Furthermore, projected impacts were presented in terms of the DALY, a standardized epidemiological framework used within the fast-growing field of 'global health' (Department of Health, 2007) for impact assessment and prediction (Stein et al, 2005).

For the CGIAR, however, biofortification research involves the conceptualization of a different type of 'user'. Farmers, the traditional target group for the CGIAR, become intermediaries in a chain that reaches out to consumers as the end-users (though there is a degree of overlap in the subsistence farmer). While the generalization inherent in categorizations such as 'small farmer' are not without problems, the location and identification of 'poor consumers', or even poor rural consumers, is far more problematic. Yet more problematic is the identification of the nutritional requirements of particular populations, which are invariably shaped by interacting local biological, socio-economic and cultural dynamics (Calloway, 1995).

Bridging agriculture and nutrition therefore involves not only a meeting of scientific disciplines, but very different approaches to defining target groups, needs and risks. As discussed in Chapter 1, this is in a context of nutrition as a field of research and practice characterized by constant questioning and revision of current knowledge and guidelines about human micronutrient requirements (Sommer and Davidson, 2002), their availability from different sources (IVACG, 2003), as well as the dynamics of inter-actions among micronutrients and between micronutrients and a range of promoters and anti-nutrients present in different food items. In this case, settling on a set of plant-breeding targets that would translate into the types of global health 'impacts' envisaged through the MDG framework is not without problems.

Given these significant challenges, it is perhaps surprising that an initiative aiming to bridge agriculture and nutrition should not only have been selected by the CGIAR Science Council as one of the four pilot Challenge Programs, but has come to be regarded as the most successful. Surely the challenges inherent in the enterprise suggest a need for greater attention to the role of local contexts in shaping human nutritional and health status in different environments, an imperative that would be at odds with the 'international public goods principle' enshrined within the Challenge Program model? The way in which these apparent contradictions have been resolved is key to understanding the significance of current developments in international biofortification research.

As this research was concluding, the sole output of rice biofortification research, in terms of varieties on the market, was MS13, the iron rice variety developed by IRRI and used as the experimental material for the iron rice feeding trial under the ADB-funded programme (2001–3). As discussed in Chapter 2, MS13 was approved and released by the Philippine National Varietal Improvement Group as a 'special variety'; however, its subsequent dissemination and adoption had been limited. In comparison, as discussed in Chapter 3, nutritional analysis of Golden Rice was still ongoing, and field trials began in the Philippines in April 2008, with an envisaged release date of 2011. Beyond the rice crop component of HarvestPlus, the main success story is the orange sweet potato project inherited from earlier initiatives. These limited, highly qualified successes must surely be a disappointment given the projections, back in the mid-1990s, of 'nutrient-enriched crops ready for commercialisation in 6–10 years' (Cribb, 1995).

A key element in the construction of the HarvestPlus enterprise has been its construction of the user around an imagined product. This involved extending a logic already established within the Golden Rice project, for which, as discussed in Chapter 3, a pathway had been constructed as the negotiation of a series of 'hurdles' (Potrykus, 2001, p1159) or 'roadblocks'.[12] In this case the two ends of this chain – the product and its associated benefits and risks, and the needs, desires and practices of the people for whom the product is intended – were black-boxed and removed from the discussion, which then focused on the impediments presented by particular roadblocks that charted the road between them. What initially appears as a ground-breaking initiative is revealed, therefore, upon closer examination, as a reproduction of an all too familiar model of linear technology transfer (Rogers, 2003).

Rather than understand the user as both an individual with particular tastes and desires and a member of social groups and networks with cultures and practices linking food with aspects of cultural, social and spiritual life, as well as health, s/he is reconstructed in terms of – another ideal type – the rational consumer. With this as the starting point, it is assumed that, given the right incentives, the rational user might be induced to 'switch' from 'non-biofortified' to 'biofortified' rice, a choice that will lead automatically to health benefits (Zimmermann and Qaim, 2004; Stein et al, 2005).

This one-dimensional construction of the rice consumer has been complemented by the creation of new categories of 'population', defined in terms of their exposure to the risk of contracting infectious diseases associated with micronutrient malnutrition (Bouis, 2004), in comparison to an abstract generic, global target. These passive, abstracted populations stand in contrast to the specific, located groups to which Calloway (1995) argued an initiative such as biofortification would need to respond in order to be effective. However, while the iron rice study represented an attempt at a more grounded, context-responsive approach, the scaling-up of biofortification to a global Challenge

Program involved a major reorientation away from these more contextualized elements, towards a more centralized approach that necessitated a closure of debates around the question of impact.

In this case, impact analysis within HarvestPlus took its cue not from the grounded lessons of the iron rice study, but from a worldview already substantially entrenched within the Golden Rice project. Despite its conscious departure from a single product focus of the Golden Rice project, notions of diffusion and impact have evolved from the influential work of Zimmermann and Qaim (2004) in constructing the socio-economic case for Golden Rice as a starting point. In doing so, HarvestPlus has kept the black box containing the complexities and uncertainties around 'impact' firmly shut.

These framings of impact, far removed from the socio-economic, cultural and ecological contexts of imagined target groups, reflect broader trends in development thinking. Rhetorical moves to universalize and globalize definitions and measures of impact speak to notions of global health and development that are narrowed to the economic area and reduced to measurable units (as exemplified by the DALY) and targets that can be isolated and replicated (Barker and Green, 1996; Murray and Lopez, 1997; Cohen, 2000). It is in this context that notions of impact that emerge from a marriage between IPGs and MDGs find acceptance.

GM or Not GM – Is that the Question?

Challenge Programs were highlighted as a point of departure for the CGIAR system, as a new *modus operandi* that would enable the system to leverage funds and access technologies from new sources, while maintaining its mandate to generate international public goods (Science Council and CGIAR Secretariat, 2004; IFPRI, 2005). This shift converged with donor trends towards more streamlined operations and the disbursal of fewer, larger grants.[13] In this context, multi-institutional, interdisciplinary research networks that appear, at least on paper, to have all bases covered became attractive both to donors and to those concerned with securing the future of the CGIAR system.

In the case of both HarvestPlus and the ProVitaMinRice Consortium (PVMRC), 'advanced research institutions' in the North played a key role, reinforcing a set of relations that became established during the Rockefeller Foundation-funded IPRB. However, Knight (2003) has drawn attention to a gradual attrition in public funding for agricultural research in general, and plant breeding in particular, in universities in the North, pressures that are only starting to be felt within the CGIAR system. He draws attention to the ways in which this decline, coupled with developments in the intellectual property environment, have taken their toll on the public infrastructure for crop research, particularly plant breeding. In its place, public research institutions have shifted their attention and limited resources to molecular biology – a surer route to publications – and plant breeding 'while still alive in the private sector ... [now works] to subtly different ends' (Knight, 2003, p569).

It is in light of these global shifts that we return to the question of interdisciplinarity in international collaborative research, and consider how both enduring and newly emerging interdisciplinary asymmetries and struggles are being enacted on this shifting terrain. The early struggles between crop scientists (or more specifically, plant breeders) and nutritionists can be seen in terms of a loosely organized field of research and practice, in which debate, dissent and attention to context are everyday aspects of professional practice, interacted with plant breeding as a discipline accustomed to its place at the apex of an established 'classic cluster' of crop sciences reflecting equally well-established academic disciplines. These contrasting 'machineries of knowing' (Knorr-Cetina, 1999, p2) met on highly unequal terms in the context of a centralized, CGIAR-based initiative that had already defined its task in terms of the 'simple and efficient criteria' of 'breeding up' grain nutrient content.

This orientation has carried over into more recent discussions at IRRI, concerning whether high-iron rice will need to 'go transgenic', in order to meet preset targets for grain nutrient content. At the outset, promoters of HarvestPlus pre-empted concerns about genetically modified crops by stressing that the programme would focus, at least in its first phase, on conventional plant breeding – while retaining the option to explore the potential of transgenic techniques in later phases. Now, it is argued, with 'proof of concept' secured from the bioefficacy study (Haas et al, 2005), there is a case for taking the next step and exploring the possibility that a transgenic approach will generate higher levels of iron content, meeting the levels required for a significant biological impact.

This neat logic obscures an area of uncertainty that continues to follow iron rice research: that of GxE (gene by environment) interaction. As discussed in Chapter 2, agronomic testing of the high-iron rice germplasm (IR68144) revealed unexpected levels of variation in both grain iron content and yield. These findings raised questions about the breeding process itself – should agronomic tests have been conducted earlier, to inform the choice of breeding parents?[14] Or, to paraphrase Simmonds (1991), was this an indication that future iron rice research should seek to exploit, rather than simply observe (and hope to minimize), these GxE dynamics?

Furthermore, preparations for the bioefficacy study had revealed the critical role of post-harvest processing, in particular the processes of milling and polishing, in determining grain iron content (Gregorio et al, 2003). Would the promotion of brown rice for its superior iron content provide a more viable way forward? While not without its difficulties, for example the problems of storage and cooking time, the exploration of alternative milling strategies as a way to avoid nutrient losses was proposed as a possible research avenue.[15] However, as experience in the Philippines has shown, rice milling is highly decentralized,[16] and therefore the outcomes of this type of research would be unlikely to generate the type of standardized, predictable outcomes that HarvestPlus is under pressure to generate.

As discussed in Chapter 2, the shift from a field-oriented Filipino 'family' of researchers to an international research network brought with it an epistemic shift from principles of interdisciplinary holism to a reductive approach in which grain iron content was identified as the key variable. As a consequence, both conventional plant breeding and transgenic research have moved in the same direction along the scale, from the site-specific to generic technologies (Biggs and Clay, 1981). This shift reinforces the pre-existing predisposition of the CGIAR system towards reductive, genetics-led approaches.

With the dispersal of the iron rice family, and the narrowing of focus to the question of iron content, the debate has polarized around the question of whether iron rice will ultimately – inevitably – have to go transgenic. Biofortification has come a long way since it was, in the early days, conceptualized as a matter of 'tailoring the plant to fit the soil' (Bouis, 1996b, p5). Crucially, one of the attractions of a transgenic approach is that it appears to obviate further investigation of the uncertainties around GxE interactions and post-harvest losses. In this way, the transgenic approach presents itself as offering the kind of predictability and controllability that such a centralized programme demands.

Boundary Terms and 'Escape Hatches'

As discussed throughout this book, CGIAR Challenge Programs such as HarvestPlus represented a shift upstream (towards basic research) and an imperative to prepare for more intense engagement in public-private partnerships. These two directions are presented within CGIAR communications as complementary, in ways that call for a more centralized structure and approach (IFPRI, 2005; Science Council, 2006). Central to these arguments is a reassertion of one of the sacred cows of the CGIAR system, the notion of research outputs as international public goods. This re-emphasis on basic research is seen as essential to a return to the 'stricter application' of the international public goods principle that will ultimately lead to impact (Science Council, 2006, p7).

Was this a point of departure for biofortification research in the CGIAR? While its framing in terms of an updated international public goods model reflected changes taking place within the CGIAR at a particular time, this was not inconsistent with more implicit processes of framing and narrowing that had been at work since the early days of the more modest CGIAR micronutrients initiative of the 1990s. In particular, the construction of a 'win–win' argument for the biofortification of staple crops with the trace minerals, iron and zinc, involved, crucially, a reframing of the job at hand from 'tailoring the plant to fit the soil' to the 'simple and efficient' method for screening for 'the micronutrient content of the seed' (Graham and Welch, 1996, p55).

By summoning the foundational CGIAR myth that desired benefits can be built into the seed and thus can be immune to the effects of institutional factors encountered downstream, early promoters were implicitly mobilizing the notion of research outputs as international public goods. This rhetorical move effectively reduced the complex dynamics of plant-soil interactions and the uncertainties around links between plant and human nutrition to the single measurement of grain nutrient content, enabling its 'deficiency' to be constructed as an 'isolable problem' amenable to genetic improvement, mirroring definitions of problem and solution around 'yield' underpinning the Green Revolution (Anderson et al, 1991. p32).

Has the CGIAR come full circle, therefore, following a succession of attempts at more context-responsive approaches, in response to critiques of the Green Revolution (Oasa, 1987)? Under pressure to demonstrate its relevance and contribution to a MDG-driven development agenda, explicit recourse to the international public goods principle presents itself as a classic 'escape hatch' (Clay and Schaffer, 1984b, p192). In this case science policy strategists driving changes within the CGIAR are able both to reassert a role for 'neutral' agricultural science and to distance themselves from downstream consequences arising from the contradictions inherent in such a notion (Oasa, 1987).

In this respect, the boundary term, 'proof of concept' has played an important role. Its transition from the business development lexicon to the world of scientific research occurred in the context of a felt need to defend Golden Rice in the face of unexpected levels of criticism. By employing this term, the Golden Rice inventors were able to shift the goalposts for success. Securing 'proof of concept' simply demonstrated that the technology 'worked' in the narrowest of terms – in this case that it was possible to engineer a biosynthetic beta-carotene pathway in rice. Pertinent questions about the contested nutritional benefits, the social and cultural acceptance of yellow rice or the food safety and regulatory implications were thus framed out of a less demanding criterion for success.

In employing this term in answer to critics, the Golden Rice inventors and promoters were able to 'top and tail' the debate, simultaneously excluding upstream framing assumptions and downstream uncertainties from the discussion. Furthermore, its apparently neutral, technical character served to obscure its inherently political function in advancing a course of action that followed from a particular set of framing assumptions, while avoiding an open discussion about the validity of those assumptions. In this context, the use of the term 'proof of concept', while borrowed from the private sector, resonates with established traditions in public policy analysis of translating 'inherently normative, political and social issues into technically defined ends' (Fischer, 2003, p131).

HarvestPlus was successful in leveraging new funds, in ways that other Challenge Programs were not, in this case from the Bill and Melinda Gates Foundation (BMGF), whose role in underwriting the extension of global biofortification research networks has been critical. Moreover, their support continued, despite the lack of results and unease expressed by other donors (see Chapter 4). This connection led, in turn, to the foundation launching its own biofortification 'Grand Challenge', facilitating the further extension of networks around biofortification. Over time, these networks and their respective agendas converged and (starting with Golden Rice) increasingly overlapped, raising questions about the future 'niche' of HarvestPlus. At the same time, the independent governance structures established for Challenge Programs created alternative routes through which new partners can access the unique combination of knowledge and resources that the CGIAR represents. In so doing, the actor-networks extending around HarvestPlus were carving out new spaces, neither public nor private.

This book has thrown new light on the tensions involved in maintaining and extending these networks; tensions which are constraining opportunities to foster genuine 'partnership' relations, engage in open interdisciplinary enquiry and respond to downstream realities. Under intensified pressures to present a certain face, programmes such as HarvestPlus follow what seems to be the path of least resistance, simplifying and black-boxing such realities and, instead, identifying their 'target' as generic 'end-users' and 'populations at risk'. In this way, programmes like HarvetsPlus reinvent the CGIAR centre as a new kind of 'centre of calculation' (Latour, 1987), reconstructing 'use', 'need' and 'impact' in ways that promise to hold together the fragile networks that point to new possibilities for the reconfiguration of international crop research.

Such developments raise pertinent questions about the governance of science and technology for development. Chapters 3 and 4 revealed how, under these increasing pressures, the role of CGIAR centres slips from broker to gatekeeper. On whose behalf do CGIAR centres find themselves playing such a role? The chapter on Golden Rice highlights a set of dynamics through which key actors have conceded access, in the narrowest of senses, to predetermined technologies, while maintaining tight control of the research process and the parameters within which it may be discussed among research 'partners', as well as in the public sphere. Such dynamics are sanctioning a return to more top-down modes of development, as well as restricting the space for scientists to share and debate findings and unresolved technical questions, in what is still a young science.

As biofortification research has become increasingly 'global', attention has shifted upstream, relocating the locus of decision-making ever further from the beneficiary groups in whose name such substantial investments are being made. In particular, the centralized design of HarvestPlus has served to narrow debates about technology choice to bipolar questions, such as whether biofortification research will have to 'go transgenic' in order to meet a set of generic targets and whether, ultimately, consumers will be induced to 'switch' from non-biofortified to biofortified varieties. A sense of urgency pervades these initiatives, as the race is on – once again – to find the universal fix (Leach and Scoones, 2006). These simplified, polarizing frames implicitly curtail discussions about what are, or should be, the goals of research, who should decide those goals, and for whom.

Conclusion

Expectations of HarvestPlus were high from the outset. In its first phase, the programme was to provide 'proof of concept' for biofortification and serve as an exemplar of global public goods science. In 2008, the influential group of economists behind the 'Copenhagen Consensus' met a second time and now threw their weight unreservedly behind biofortification, declaring it to be one of the 'top five solutions' to the world's biggest problems.[1] Later that year, a new phase of reforms was announced at the CGIAR AGM in Maputo, Mozambique, signalling a shift from 'proof of concept' studies to the 'scaling up' of proven 'best bet' technologies and programmes. By streamlining activities around a series of 'best bets', CGIAR would work towards a portfolio of 'mega-programmes' that would meet its strategic objectives (von Braun et al, 2008). Biofortification was included on this list. In the process, it underwent an apparently seamless transition from still ongoing 'proof of concept' enquiry to 'best bet' status, ready for 'scaling up'.

In January 2009 funding was approved for a second phase of HarvestPlus.[2] Meanwhile, the onset of a global food crisis had, promoters argued, strengthened the case for 'scaling up biofortification' yet further:'"As food prices continue to rise, and people are forced to reduce their consumption of more nutritious foods, such as animal products and leafy vegetables, micronutrient malnutrition will increase" [Howarth Bouis, Director of HarvestPlus] cautioned. "Biofortification will therefore become all the more important as a strategy to improve nutrition and health."';[3] 'Because of a condition called micronutrient malnutrition, the global food crisis, even if it eases soon, will have repercussions for decades ... "Biofortification is really the last hope for the poorest of the poor," says Fil Randazzo, a senior program officer at the Bill and Melinda Gates Foundation, which has contributed $38 million to HarvestPlus over the last five years.'[4]

These developments represent a continued commitment to a goal-oriented approach to research and development, which resonates with a contemporary international development agenda defined and delimited by the MDG framework and a convergence of the fields of agriculture, nutrition and health around technical, central and reductive approaches. In this case, the proposed 'mega-programmes' both sanction a reassertion of the CGIAR's foundational myths and promise the kind of generic, scaleable 'silver bullet' technologies that satisfy prevailing visions of 'impact at scale'.

Focusing on 'the world's most important crop', rice, this book has explored the case of biofortification research as an exemplar of science-policy processes in international crop research. Singled out as an early pilot for the CGIAR Challenge Program model and the first tranche of the BMGF's Grand Challenges in Global Health, international biofortification research seemed to offer exciting new possibilities in cutting-edge, interdisciplinary, collaborative enquiry. In practice, biofortification research has, from its modest and diverse beginnings, converged on a very singular pathway. Over a period of more than ten years, a range of possible directions in biofortification research has been successively framed out of a debate that now seems to point in just one direction. It seems likely that other 'best bets', such as those aiming to develop drought-tolerant maize for Africa, for example, will follow a similar pattern (Brooks et al, 2009).

This book has traced some of the processes of closure that have brought biofortification research to this point. In the first place, the framing of biofortification as a solution embedded 'in the seed' reflected an inherited pattern of interdisciplinary relations that is an enduring legacy of the Green Revolution (Anderson et al, 1991). These relations both reflect and reinforce a still dominant linear science policy model (Rogers, 2003) that continues to draw a boundary between 'upstream' and 'downstream' worlds; 'strategic' and 'adaptive' research; and 'innovation' and subsequent 'diffusion'.

These interdisciplinary hierarchies and linear models, long 'established' (Clay and Schaffer, 1984a) in particular sets of intra- and inter-institutional arrangements, particularly between CGIAR centres and NARS, have remained, despite successive institutional reforms, remarkably resistant to change over the years (Oasa, 1987; Eicher and Rukuni, 2003). In particular the notion of the 'mega-programme', now central to CGIAR future plans, reinforces an established division of labour in agricultural research in which international centres produce generic technologies for adaptation to specific environments and user-groups by national 'partners'. As such, it reinforces a model of upstream–downstream relations that allows actors located upstream to maintain a 'blind spot' that shields them from downstream complexities, uncertainties and surprises.

A key conclusion of this book is that a consolidation of linear, centralizing blueprints for science and development threatens to close down yet further the spaces for scientists and others to 'look sideways' and consider, in a more grounded and disaggregated fashion, the expected and unexpected ways in which outputs of their research might interact with diverse contexts. There is, it seems, no room for doubt. Of equal concern is the way in which tenuous results have been transformed into a solid foundation for 'scaling up'. An atmosphere of urgency continues to pervade these initiatives, exerting a disciplinary force that marginalizes more questioning voices. The question then, is what kind of institutional changes would open up these spaces to allow the consideration of uncertainties and alternatives. The following paragraphs attempt to answer this question.

Locating and Engaging 'Users'

The CGIAR and its partners, supported by new donors, notably the BMGF, are tackling increasingly complex global problems. This complexity is reflected in institutional arrangements that link upstream actors from a variety of sectors and disciplines, but stop short of engaging with the diversity of contexts from which such complexity ultimately derives. Rather than respond to diversity and complexity, such programmes construct alternative realities more attuned to the globalized configurations it has taken so much effort to build. This book has provided a glimpse of the gulf that exists between these constructed worlds and the lived realities of projected beneficiaries of these programmes.

In particular, questions of user engagement and participation have been supplanted by notions of 'consumer choice', which frame the provision of new technologies in terms of extending the choice available to the poor. This framing deftly steers debates surrounding agricultural research and innovation away from power–knowledge relations between farmers and scientists. 'Users' are now cast as consumers of technologies already in the pipeline. In the process, diverse, complex needs are transformed into demand for predetermined products, completing the circle that reaffirms, for upstream actors and donors, an emerging *modus operandi* for global, public goods science.

Notably missing from these deliberations are the developing-country governments, line ministries and research bodies who might otherwise have something to say about national and sub-national priorities and problems, and are instead cast as the first line of beneficiaries of international funding and global research outputs. The case of iron rice in the Philippines demonstrated that the ways in which biofortification pathways mapped on to the broader canvas of national priorities and policy choices surrounding food, nutrition and agriculture could not have been predicted from a distance. This is where the alternative worlds constructed by programmes like HarvestPlus, in which diverse realities are 'aggregated up' into potential win–win scenarios, diverge most sharply from the everyday realities of resolving trade-offs between competing priorities. As one observer asked: 'Where are the voices of people HarvestPlus is supposed to be targeting? Why, if I'm a health minister, should I put money into this?'[5]

These lessons of this book indicate that questions of who and where – which people, in what place – should precede, not follow, the setting of research targets if they are to respond to the needs of 'people in places'. To address these lessons requires a radical rethink of an upstream–downstream model that continues to structure the international division of labour in crop research. New models are needed which provide space for those local institutions best positioned to understand and engage with diversity as it exists in different localities (Anderson et al, 1991), rather than reconstruct it as a sequence of hurdles that stand between the products of 'global science' and its projected beneficiaries.

Rethinking Upstream–Downstream Relations

Challenge and mega-programme models rely on a widely accepted binary that separates 'upstream' activities such as basic research from 'downstream' adaptation, dissemination and adoption, following a classic linear innovation model (Rogers, 2003) whose underlying assumptions continue to guide science policy thinking inside and outside the CGIAR. These prescribe the role of 'partners' located downstream as being to facilitate the transfer of research outputs (the value or appropriateness of which they need not question) along a technology 'pipeline' towards the 'end-user'.

At the same time, while these downstream actors are welcome to debate issues of operationality – how to maximize efficiency, pace and scale – participation in debates about directionality – which research goals should be pursued and why – is restricted to groups of appointed 'experts' such as a Science Council (for the CGIAR) or Scientific Board (for the BMGF). This separation of directionality and operationality allows those making 'strategic' decisions about research directions to avoid public accountability for their choices. Meanwhile, questions about the distribution of risks and benefits among and between 'populations' are relegated to a tactical level, to be dealt with later, once products move further 'downstream' (Brooks et al, 2009a).

The biofortification pathways outlined in this book have diverged in significant ways, reflecting alternative framings of the problem at hand. A key lesson is that these pathways are inseparable from the social, institutional and geographical location of the networks of scientists and other actors that form around them. Issues of directionality, therefore, are inextricably linked with the structure of research partnerships and the power-knowledge dynamics that shape practice within scientific institutions and research networks. This goes beyond the question of whether such partnerships are technology-driven or user-driven (Ayele et al, 2006). A more open discussion about the relative merits of alternative biofortification pathways requires more fundamental changes to the 'rules of the game' that prescribe the role of actors according to their 'upstream' or 'downstream' location, and the types of partnerships that are, as a consequence, forged between them.

Towards a More Reflexive 'Public Goods' Science?

These proposals present a fundamental challenge to a CGIAR system that has resisted pressures to restructure its relations with other institutions, particularly at the national and local level. As this book has highlighted, these external relations cannot be separated from an internal interdisciplinary ordering based on embedded assumptions about the value of certain types of knowledge over others. Perversely, the challenges of interacting with a more heterogeneous set of partners appear to have led, not to greater openness to new ways of knowing, but to a defensive retreat into foundational myths and outdated practices.

This study has revealed that securing consensus upstream does not necessarily lead to integrated science practice downstream. An alternative approach would be one that builds consensus between upstream and downstream actors. While a departure from current practice, such an approach would not be without precedent. In 1999, IRRI hosted a workshop that showcased the findings of the CGIAR Micronutrients Project (1994–9) and provided a platform for crop and human nutrition scientists to openly debate some of the upstream *and* downstream challenges involved in 'improving human nutrition through agriculture'.[6]

Perhaps the time has come to reconvene a similar gathering. This would provide an entry point for building a different type of consensus – rather than rely on a global protocol, itself the outcome of a prior 'consensus' that has emerged from negotiations and struggles between actors located in different places and times. This might, for example, open space for a more grounded integration of the perspectives of crop scientists more attuned to the environmental factors that have continued to thwart plant breeders' efforts to meet HarvestPlus targets, as well as those of nutritionists more knowledgeable about the geographical, socio-economic, cultural and seasonal variations in the 'typical Filipino diet'.

CGIAR reforms launched in 2008 in pursuit of a 'revitalized CGIAR' included a commitment to create an 'exciting research environment ... that will support great science'.[7] This book has shown how an emerging practice of goal-oriented research and development frustrates such aims, restricting open, rigorous scientific enquiry. Could biofortification research become an exemplar of a different kind? As an inherently boundary-crossing, interdisciplinary (or even trans-disciplinary) field it presents an opportunity to acknowledge these constraints, move beyond the rhetoric and take genuine steps to foster such an environment.

Notes

Introduction

1. Debates about the consequences of the Green Revolution and its aftermath are discussed in Chapter 1, which outlines the historical development of international agricultural research and the institutional infrastructure (in particular, the CGIAR) that evolved to support it.
2. For a recent initiative attempting to address this imbalance see: www.salzburgseminar.org/2008/aai.cfm (8 June 2008).
3. www.worldfoodprize.org/assets/symposium/2005/transcripts/Bouis.pdf (9 January 2009).
4. In its first phase (2003–7), the HarvestPlus programme focuses on six crops – rice, wheat, maize, cassava, sweet potato and common bean – and three micronutrients – pro-vitamin A, iron and zinc.
5. Interview, IRRI, 24 May 2006.
6. This programme, the International Program on Rice Biotechnology (IPRB), is discussed in more detail in Chapter 3.
7. The origins of the employment of this term in relation to biofortification research, which began with the Golden Rice project, are discussed in Chapter 3.
8. Interview, HarvestPlus, 27 January 2006.
9. www.gmwatch.org/profile1.asp?PrId=294 (18 March 2008).
10. www.genecampaign.org/News/golden-rice.htm (18 March 2008).
11. This phrase encapsulates a set of ideas about the historical location of IRRI, discussed in Chapter 1 (see Cullather, 2004).
12. Here 'Cornell' refers to the location of Cornell University campus, where relevant faculties (agriculture, nutrition) as well as the USDA/ARS Federal PSNL laboratory are based.
13. G. E. Marcus, keynote speech at a workshop entitled: 'Problems and Possibilities in Multi-sited Ethnography', University of Sussex, UK, 27–28 June 2005.
14. Interview, Washington, DC, 25 January 2006.
15. The circulation to informants of a summary of research findings and/or the doctoral thesis (Brooks, 2008) prompted some debate, but these were exceptional instances that did not, for the most part, involve the more central protagonists.
16. For example, see www.scidev.nnet/en/news/uk-to-streamline-health-aid-strategy.html (8 April 2008).

Chapter 1

1. Title borrowed from Hawkes and Ruel (2006a) on agriculture-health linkages.
2. www.ifpri.org/pubs/newsletters/ifpriforum/200306/if02biofort.htm (18 March 2008).
3. J. George Harrar and Norman Borlaug, who each played a key role in the Green Revolution, were respectively the director and chief scientist of this programme.
4. Specifically, this was 'discussed before the US House of Representatives at the Subcommittee on National Security Policy and Scientific Development, of the Committee on Foreign Affairs'. The title given to the publication of the proceedings was *Symposium on Science and Foreign Policy: The Green Revolution'* (see Spitz, 1987, p56).
5. This 'technology' definition requires qualification, however. The tendency to describe Green Revolution technology in terms of HYVs alone has resulted in misleading claims for its 'scale neutrality'. In fact, it was a technology 'package' which required farmers to make several concurrent changes if they were to produce the 'optimal conditions' necessary to achieve the stated yields, a feature which led to a strategy of targeting 'progressive farmers' (Pearse, 1980, p181; Glaeser, 1987, pp1–2).
6. Later studies disputed these findings, however (for example, see Lipton and Longhurst, 1989; David and Otsuka, 1993).
7. FSR is a broader term than cropping systems research, which encompasses both multiple crops and livestock.
8. 'First Review of Systemwide Programmes with an Ecoregional Approach', available at: www.sciencecouncil.cgiar.org/publications/html/ X5783E/X5783E00.htm (25 January 2008).
9. www.grain.org/seedling/?id=115 (18 March 2008).
10. In 1987 IRRI was investigated in the Philippine Senate for 'abusing its diplomatic privileges and importing foreign isolates without import permits'. These events prompted the development of the world's first national biosafety framework (Perlas and Vellvé, 1997, pp164–5) several years before international negotiations led to the ratification of the Cartagena Protocol.
11. The question of why it took so long for the Philippines to establish its own NARS is a contentious one, which highlights IRRI's ambiguous location 'in but not of the Philippines' (Cullather, 2004). IRRI has suggested that the Philippines government had 'misconstrued IRRI's mission' by assuming it would 'substitute for a national rice research programme'. However, from their analysis of archive material, Perlas and Vellvé conclude that IRRI had 'persuaded senior officials that the institute would do such a good job that research on rice by the government of the Philippines was not urgent' (1997, p30).
12. www.grain.org/seedling/?id=389 (18 March 2008).
13. www.fao.org/Wairdocs/TAC/X5796E/x5796e03.htm (4 May 2007).
14. www.grain.org/seedling/?id=39&print=yes (22 January 2008).
15. www.irri.org/about/images/Important%20DTES%20IRRI%20History,% 201959-2006.pdf (24th January 2008).
16. www.sciencecouncil.cgiar.org/publications/pdf/SCC1005.pdf (18 March 2008, emphasis added).
17. Quoted at www.cgiarnews.org/enews/june2006/story_03_print.html (25 July 2007, emphasis added).
18. www.grain.org/seedling/?id=389 (18 March 2008); www.fao.org/Wairdocs/TAC/ X5796E/x5796e03.htm (4 May 2007).

19. The main UN institutions responsible for nutrition are UNICEF, WHO, FAO and WFP. These institutions convene the UN Standing Committee on Nutrition (SCN), which oversees a diversity of working groups on different aspects of international nutrition. In total, more than 14 UN institutions, 5 international and regional development banks, over 30 INGOs, more than 20 universities and research centres, 15 CGIAR centres, several hundred academic journals and at least 12 multinational companies are working, in some way, to reduce undernutrition (Morris et al, 2008).

20. http://unstats.un.org/unsd/mi/pdf/MDG%20Book.pdf (18 March 2008).

21. The protein era was the longest to date in international nutrition, lasting from the 1930s into its gradual decline though the 1960s and 1970s (Gillespie et al, 2004).

22. These included the World Summit for Children in 1990, the Montreal Conference on Hidden Hunger in 1991 and the International Conference on Nutrition in 1992 (Solon, 2000, p515).

23. 'The World Bank is now the world's largest external funder of health, committing more than $1billion annually in new lending to improve health, nutrition and population in developing countries' (Prah Ruger, 2005).

24. www.who.int/nutrition/topics/ida/en/index.html (18 March 2008).

25. WHO interview, 14 March 2006.

26. www.fao.org/righttofood/principles_en.htm (18 March 2008).

27. www.micronutrient.org.home.asp (18 March 2008).

28. Interview, HarvestPlus, 27 January 2006.

29. See Chapter 2 for a more detailed discussion of the role of PSNL scientists in international biofortification research.

30. Funding was provided by USAID and Danish International Development Agency (DANIDA).

31. See *Food and Nutrition Bulletin*, Special Issue, November 2000.

32. Interview, HarvestPlus, 27 January 2006.

33. The biopolitical implications of this type of reasoning are explored in Chapter 3. The term 'biopolitics' was first used by Michel Foucault (1976).

34. Other nutrition projects in the top 17 were: 'development of new agricultural technologies' (5), 'improving infant and child nutrition' (12) and 'reducing the prevalence of low birth weight' (13). See www.copenhagenconsensus.com (18 March 2008).

35. www.rockfound.org/iniatives/index.shtml (23 July 2008); www.rockfound.org/about_us/news/2007/0408philanthropy.shtml (27 July 2008).

36. Notably, the HarvestPlus project timeline anticipated the commencement of 'participatory plant breeding' during the period 'Years 5–7'.

37. www.grain.org/seedling/?id=389 (18 March 2008).

38. 'Report of the First Challenge Program Review of the HarvestPlus Challenge Program', CGIAR (2007), available at: www.cgiar.org.cn/pdf/agm07/agm07_harvestplus_cper_overview.pdf (25 January 2008).

Chapter 2

1. See Chapter 1 for a more extended discussion of the impact of three international conferences on the early 1990s. Among these was the 'Ending Hidden Hunger' conference, which focused specifically on micronutrient deficiencies. For a discussion of the impact this had on nutrition policy in the Philippines, see dela Cuadra (2000).

2. Philippines, Bangladesh, Indonesia and Vietnam: www.adb.org/documents/prf/nutrition.asp (10 November 2005).

3. See Chapter 3 for a discussion of the meaning and use of the term 'proof of concept'.

4. HarvestPlus interview, 27 January 2006.

5. HarvestPlus interview, 27 January 2006.

6. The term 'bioavailability' refers to the proportion of nutrient the body can extract from food items and make available for utilization.

7. Interview, PSNL, Cornell, 19 January 2006.

8. '"The potential economic returns on research aimed at helping Turkish farmers on zinc-deficient lands reduce their seeding rate are tremendous,' says Braun [of CIMMYT's Wheat Programme]. 'A reduction of 80kg/ha could save about 400,000 tons of seed a year, with an estimated value of US$80 million"' (CIMMYT, 1995).

9. Welch has consistently cautioned against a breeding strategy of reducing anti-nutrients such as phytates, as these have other important functions for plant growth and possibly also for human health, for example as anti-carcinogens. See Graham and Welch (1996); Welch (1996); Welch and Graham (2002).

10. http://query.nytimes.com/gst/fullpage.html?sec=health&res=990CE7 D61131F93AA15752C0A963958260&fta=y (30 May 2008).

11. HarvestPlus interview, 27 January 2006.

12. Senadhira was awarded Sri Lanka President's Award for Scientific Achievement in 1981 and the Ceres Medal from FAO in 1982. In 1998 he was awarded the Fukui International Koshihikari Rice Prize 'in recognition of his outstanding achievements in developing improved rice varieties' (Khush, 1998, p7).

13. Dr Senadhira passed away in 1998. This account of the origins and early stages of iron rice research at IRRI is based on interviews with his former colleagues and published material.

14. Interview, IRRI scientist, 9 June 2006.

15. The full name of the variety, normally abbreviated to IR68144.

16. Rats were used in preference to pigs, which are normally preferred for testing iron bioavailability (King, 2002, p512s), since this would have been costly at the screening stage (Welch et al, 2000).

17. An account of the long history of rice fortification in the Philippines, co-authored by Drs Rudolf Florentino and Ma Regina Pedro, both formerly with the FNRI, can be found at: www.unu.edu/unupress/food/V192e/ch09.htm (18 March 2008).

18. Interview, FNRI, 21 June 2006.

19. This has been accompanied by the emergence of a new phenomenon in urban centres known as 'surrogate *ulam*', where poor people add flavour to rice by adding the 'new viands' such as 'coffee, pork oil, brown sugar and Pepsi' (Datinguinoo, 2005, p88).

20. Interview, IRRI, 24 May 2006.

21. Interview, nutritionist, Manila, 22 June 2006.

22. In the Philippines, a meal without rice, however substantial, is considered just *merienda*, a snack. Even McDonalds has adapted, by including the 'McRiceBurger' on the menu.

23. Interview, FNRI, 21 June 2006.

24. 'Bioefficacy' refers to the proportion of ingested nutrient that is metabolized into its active form through the process of 'bioconversion'. www.unu.edu/unupress/fppd/V192e/ch09.htm (18 March 2008).

25. http://doh.gov.ph/foodfortification/downloads/SPSProgram.pdf (18 March 2008).
26. Interview, NFA, 25 January 2007.
27. Interview, FNRI, 21 June 2006.
28. Interview, NFA, 25 January 2007.
29. Interview, IHNF, UPLB, 26 May 2006.
30. Interview, IHNF, UPLB, 26 May 2006.
31. HarvestPlus interview, 27 January 2006.
32. www.adb.org/documents/prf/nutrition.asp (10 November 2005).
33. IRRI interview, 29 May 2006.
34. Interview, IHNF, UPLB, 7 June 2006.
35. Interviews, IHNF, UPLB, 26 May 2006; IRRI, 9 June 2006.
36. Narciso left the project after the initial setting-up period to pursue a master's degree in the United States. Sison transferred to IFPRI soon after the commencement of the feeding trial, but she was involved in the preparatory production, milling and cooking studies of IR68144.
37. Dr Del Mundo passed away in 2004, just as the researchers were completing the analysis for publication (Haas et al, 2005, p2830).
38. http://casualsavant.typepad.com/photos/fri/friends/index.html (18 March 2008).
39. The idea of a family having members scattered around the globe is, of course, a characteristic common to many Filipino families. Government of the Philippines statistics show that: 'The total number of OFWs [overseas foreign workers] deployed in 197 country destinations in 2006 hit a historic-high of 1,062,567' (www.poea.gov.ph/stats/2006Stats.pdf, 9 April 2008). This is from a total population estimated at 90.4 million (www.census.gov.ph, 9 April 2008).
40. Interview, IHNF, UPLB, 7 June 2006.
41. The role of the Catholic Church in Philippine politics and social development can be difficult for outsiders to understand, since it combines extreme conservatism – for example divorce remains illegal in the Philippines – with radical, emancipatory politics: The decision of the then head of the Catholic Church in the Philippines, Cardinal Jaime Sin, to call people out on to the streets in 1987 was a turning point in the fortunes of the 'people power' movement that ended martial law and swept Corazon Aquino to power, that year.
42. See a recent press release on the HarvestPlus website, referring to Bouis as 'Mission Man' (www.harvestplus.org/newsroom.html, 9 April 2008).
43. Interview, IHNF, UPLB, 7 June 2006 (original emphasis).
44. Discussions with 'family' members, en route to visit participating convents, 21 December 2006.
45. Discussions with 'family' members, en route to visit participating convents, 21 December 2006.
46. Interview, IRRI scientist and 'family' member, 4 December 2006.
47. Interview, Department of Agronomy, UBLB, 22 February 2007.
48. Discussions with 'family' members, en route to visit participating convents, 21 December 2006.
49. Interview, RVIG member, 17 January 2007.
50. Interview, IRRI scientist, 9 June 2006.
51. Interview, IRRI scientist, 30 May 2006.
52. Maligaya is the village in Nueva Ecija where the NSIC/RVIG and PhilRice are based.
53. Interviews, RVIG, 17 January 2007; IRRI, 9 June 2006.
54. William Padolina, deputy director-general for partnerships at IRRI, had suggested the variety be named 'Fe' – which is Spanish for faith, as well as the symbol for iron – but this was never followed up (Padolina, 2003, p78).

55. Rice Technical Working Group, which reports to the RVIG.
56. Interview, IRRI Scientist, 30 May 2006.
57. GxE, or 'genotype by environment', refers to the interactions between the effects of genetic factors (manipulated, for example, through plant breeding or molecular biology) and the effects of environmental factors such as soil composition and moisture content, as well as cultural practices in farming. In conventional crop research, GxE trials involve the planting of new germplasm in contrasting agro-ecological environments in order to establish whether mean performance and variance are within acceptable limits. This limited consideration of environmental factors within crop science has been criticized (notably by Simmonds, 1991). These debates are explored further in Chapter 4.
58. Interview, PhilRice scientist, 30 May 2006, original emphasis.
59. Interview, PhilRice scientist, 7 June 2006.
60. Interview, IRRI scientist, 8 June 2006.
61. One debate that has highlighted these epistemological and methodological differences between crop sciences in recent years has been the SRI (System of Rice Intensification) controversy, sometimes referred to as the 'rice wars' (Uphoff, 2004). An accumulated set of practices rather than a technology as such, SRI emphasizes the adoption and adaptation of a set of crop management practices (in other words, an emphasis on the E of GxE) which, it argues, can generate impressive yield increases without the use of genetic manipulation (Stoop and Kassam, 2006). The claim that such an approach may be comparable or even preferable to the genetics-led paradigm that has hitherto dominated international rice science in general, and IRRI in particular (Anderson et al, 1991), has provoked a strong reaction within established international rice science (Uphoff, 2004).
62. Interview, IRRI, 29 May 2006.
63. The focus, in the paper by Haas et al (2005) on cooked rice seems to have added to the confusion about iron content and bioavailability.
64. Annual award given by the National Academy of Science and Technology (NAST) of the Philippines (Gregorio, 2006).
65. Interview, PhilRice scientist, 16 January 2007.
66. Interview, RVIG member, 17 January 2007.
67. By 2007 the Philippines was the world's largest rice importer. In 2008, with global rice prices escalating and national stocks low, the government was urging the public to refrain from stockpiling rice and encouraging fast-food outlets to serve half portions of rice with each meal, as fears grew that rising prices and diminishing supplies would lead to unrest. See http://news.bbc.co.uk/1/hi/world/south_asia/7324596.stm (9 April 2008).
68. While yields were increased by approximately 13 per cent in the dry season and 9 per cent in the wet season, there was considerable variability, which may account for the slow rate of adoption (Norton and Francisco, 2006, p158).
69. Interview, Manila, 30 January 2007.
70. 'The *Nacionalista* ticket headed by Ferdinand E. Marcos had won the election in 1965 on the slogan "Progress Is a Grain of Rice"' (Cullather, 2004, p243).
71. Interview, ANGAT-Laguna, Los Baños, 7 December 2006.
72. http://sws.org.ph (2 February 2007).
73. Interviews, Philippine Rural Reconstruction Movement and Philippine Greens, 24 January 2007; Philippines Peasant Institute, 30th January 2007.
74. Interview, RVIG member, 17 January 2006.

Chapter 3

1. 'The term "transgenic crops and plants" (or "transgenics") means any plant or resulting seed manipulated by recombinant DNA technology to express a gene encoding a novel trait'. Found at www.agriculture.purdue.edu/PAC/transgenic.pdf (30 May 2008).
2. This accounted for approximately half the foundation's agricultural funding (Toenniessen, 2000, p4).
3. Leocardo Sebastien, also a Rockefeller Foundation scholar, is now director of PhilRice.
4. Important findings were the discovery of two levels of vitamin A deficiency: clinical, causing blindness, and sub-clinical, less visible but having an eroding effect on the immune system (which in turn undermines micronutrient absorption within a vicious cycle of malnutrition-infection) (Bhaskaram, 2002, pS40).
5. Other micronutrients, in particular zinc (and also folate), have since been added to this agenda, see: www.micronutrientforum.org/index.cfm (18 March 2008).
6. www.hki.org/network/Philippines.html (18 March 2008).
7. The IVACG recently merged with the International Nutritional Anaemia Consultative Group (INACG) to form the Micronutrients Forum: www.micronutrientforum.org/AboutUs.cfm (18 March 2008).
8. www.goldenrice.org/Content1-Who/who_Ingo.html (18 March 2008).
9. www.nj.com/specialprojects/index.ssf?/specialprojects/rice/rice4.html (18 March 2008).
10. www.agbioworld.org/biotech-info/topics/goldenrice/tale.html (18 March 2008).
11. Interview with a former IRRI director, London, November 2005.
12. www.nysaes.cornell.edu/comm/gmo/ag_program.pdf (18 March 2008).
13. www.nysaes.cornell.edu/comm/gmo/ag_program.pdf (18 March 2008).
14. www.time.com/time/magazine/article/0,9171,997586-1,00.html (30 May 2000).
15. www.goldenrice.org/PDFs/The_GR_Tale.pdf (20 April 2008).
16. www.seedquest.com/News/2005/may/12362.htm (19 June 2007).
17. On 2 December 1999, Syngenta AG was formed from the spin-off and merger of Zeneca Agrochemicals, the crop protection business of AstraZeneca, and Novartis's crop protection and seeds businesses. See www.syngenta.com/en/media/article.aspx?pr=061900&Lang=en (18 March 2008). Novartis had resulted from a previous merger between divisions of ICI (Imperial Chemical Industries) and Ciba (of Ciba-Geigy); AstraZeneca arose from the merger, in the previous year, between Astra AB of Sweden and the Zeneca Group PLC, UK: www.syngenta.com/en/about_syngenta/timeline.aspx (18 March 2008).
18. http://online.sfsu.edu/~rone/GEessays/goldenricehoax.html (18 March 2008).
19. This *de facto* moratorium was in place from 1998 until 2004, see www.gmo-safety.eu/en/archive/2004/289.docu.html (20 April 2008).
20. Derham at www.seedquest.com/News/2005/may/12362.htm (19 June 2007).
21. www.grain.org/briefings/?id=35 (18 March 2008).
22. Interview, ISAAA, 15 December 2006.
23. www.goldenrice.org/Content1-Who/who1_humbo.html (18 March 2008).
24. www.seedquest.com/News/2005/may/12362.htm (19 June 2007).
25. www.goldenrice.org/Content1-Who/who1_humbo.html (18 March 2008).
26. www.irri.org/media/press/press.asp?id=56 (18 March 2008).
27. Interview, IRRI, 5 December 2006.

28. Interview, IRRI, 5 December 2006.
29. IR68144 had been identified by Dharmawansa Senadhira and his team in the salinity plant breeding group as a high-iron, high-zinc variety in the early 1990s and during 2001–3 was the material used for the iron rice feeding trial in the Philippines (see Chapter 2).
30. IRRI interview, 5 December 2006.
31. The sudden departure of Swapan Datta, the lead scientist working on Golden Rice, from IRRI, at the time Barry took up his post as Golden Rice network coordinator has been a matter of some controversy. See www.genecampaign.org/News/golden-rice.htm (18 March 2008).
32. While at Monsanto Barry had been technology leader for rice, Monsanto (1995–7); core team member for High Yield Rice project (joint project with Japan Tobacco); (1995–2000); co-director for Rice Strategic Business Team, Monsanto (1997–9); head of Rice Genomics, Monsanto (1999–2001); director of research, Product and Technology Cooperation, Monsanto (2000–3). See www.irri.org/about/irridir/staffbio.asp (18 March 2008).
33. www.goldenrice.org/Content1-Who/who3_collab.html (18 March 2008).
34. Interview, IRRI, 5 December 2006.
35. GR1 and GR2 were initially distinguished from the prototype versions with the names 'Syngenta Golden Rice' 1 and 2 (SGR1 and SGR2).
36. These studies were led by Robert Russell (also a Humanitarian Board member) at the Laboratory of Human Nutrition at Tufts University. Preliminary findings were presented on a poster at the Micronutrient Forum in Turkey in April 2007. See: www.ars.usda.gov/research/publications/publications.htm? SEQ_NO_115=2105 (8 January 2008) followed by a published article the following year (Tang et al, 2009).
37. Mackintosh et al (2008) have drawn attention to the conflict between the roles of 'broker' and 'gatekeeper'. See also Chapter 4 for a more extended discussion of this issue.
38. http://syngenta.com/en/media/article.aspx?pr=101404&Lang=en (18 March 2008).
39. Interview, NARS scientist, 16 January 2007.
40. http://news.bbc.co.uk/1/hi/sci/tech/4386933.stm; www.newscientist.com/article.ns?id=dn7196 (18 March 2008).
41. Interview, IRRI scientist, 30 May 2006.
42. In contrast, this is the approach chosen for research at IRRI on transgenic iron rice (see Chapter 4). Interview, IRRI, 18 December 2007.
43. www.goldenrice.org/Content2-How/how4_regul.html (18 March 2008).
44. Epifanio de los Santos Avenue (EDSA) is the highway in Metro Manila and the location of the 'people power' revolution that saw the end of martial law, sweeping Corazon Aquino to power. This movement is now known as EDSA 1. In 2001 EDSA 2 swept Joseph (Erap) Estrada from power, then in a counter-movement, EDSA 3, his supporters attempted – and failed – to reinstate him (Bello, 2004).
45. Interview, 24 January 2007 (original emphasis). See also Perlas and Vellvé, 1997.
46. Interviews, Manila, 24 January, 1 February 2007.
47. A division of labour was therefore established in which the biosafety committee governs the steps up until contained release, at which point the DA then takes over and follows the product through to commercial release.
48. www.monsanto.com/monsanto/ag_products/input_traits/products/ yieldgard_corn_borer.asp (18 March 2008).

49. This point is arguable: there are two types of corn (maize) grown in the Philippines, yellow corn for feed and white corn for food. In practice, however, many people eat yellow corn.

50. Interview, Manila, 5 January 2007.

51. Interview, Manila, 1 February 2007 (original emphasis).

52. Interview, Manila, 24 January 2007.

53. Shiv Visvanathan quoted at www.ifpri.cgiar.org/pubs/books/IndiaProc/IndiaProc_session05.pdf (18 March 2008).

54. Interview, IRRI, 13 December 2006.

55. www.gmo-compass.org/eng/news/358.docu.html (9 June 2008).

56. Results of these bioavailability studies have recently been published (Tang et al, 2009). These studies sparked a new phase of controversy around the Golden Rice project: http://eldib.wordpress.com/2009/03/21/the-golden-rice-scandal-unfolds/ (6 August 2009) and www.tuftsdaily.com/friedman-researchers-ethics-questioned-for-feeding-children-genetically-modified-rice-1.1644770 (15 February 2010).

57. www.scidev.net/dossiers/gmcrops/gmpolicygregorio.html (7 December 2006).

58. Interview, IRRI, 11 December 2006.

59. Interview, PhilRice, 15 January 2007.

60. Interview, Manila, 5 January 2007.

61. The exclamation 'yellow rice, oh my God!', voiced by one nutritionist, indicates the challenges ahead in this respect (Interview, Manila, 22 June 2006).

62. For example see: www.monsanto.co.uk/news/ukshowlib.phtml?uid=6874 (18 March 2008).

63. Interview, IRRI, 24 May 2006.

64. Interview, IRRI scientist, 11 December 2006.

65. www.goldenrice.org/Content2-How/how8_tests.html (9 August 2006).

66. 'Grand Challenges for Global Health' initiative: Grand Challenge No.9 (GC9), to 'Create a full range of optimal, bioavailable nutrients in a plant' (four projects on rice, cassava, sorghum and banana). Rice component: 'Engineering rice for high beta-carotene, vitamin E, protein, iron and enhanced iron and zinc bioavailability' conducted by the ProVitaMinRice (PVMR) Consortium: Universities of Freiburg (Beyer), Michigan State (DellaPenna), USDA-ARS/Baylor (Grusak) and Hong Kong (Sun); and Cuu Long, Vietnam (Hoa), PhilRice (Sebastien) and IRRI (Barry). See Chapter 6 for further discussion about this initiative.

67. www.goldenrice.org/Content2-How/how10_PVMRC.html (9 June 2007).

Chapter 4

1. Interview, Cornell, 31 January 2006.

2. Interview, HarvestPlus, 17 January 2006.

3. Interview, IRRI, 20 February 2007.

4. The seminar, discussed in Chapter 2, was documented in a special edition of *Food and Nutrition Bulletin* (vol 21, no 4, 2000).

5. Interview, Science Council member, 19 January 2006.

6. Interview, Science Council member, 19 January 2006.

7. This is a recent term used within the CGIAR to refer to CGIAR centres. See: www.cgiar.org/who/structure/system/fhao/index.html (18 March 2008).

8. Advanced Research Institute: this term refers, mainly, to universities and other research institutions in the North. In practice many of the ARIs in these networks are US land-grant universities.

9. National Agricultural Research and Extension System: this is an alternative term to the more commonly used NARS, recognizing the importance of extension in agricultural innovation.

10. CGIAR quoted at: www.gmwatch.org/print-profile1.asp?PrId=295 (18 March 2008).

11. Work in Progress seminar discussion, Institute of Development Studies, University of Sussex, 4 December 2007.

12. For example, see: www.copenhagenconsensus.com/Default.aspx?ID=158 (18 March 2008).

13. Interview, HarvestPlus, 27 January 2006.

14. The IDRC is a long term supporter of micronutrient programmes, such as the Ottawa-based Micronutrients Initiative (MI): www.micronutrient.org/english/view.asp?x=1 (29 September 2009).

15. Interviews, HarvestPlus, 27 January 2006, and University of Sussex, 27 January 2006.

16. USAID was increased to US$3 million per year in 2004 (HarvestPlus, 2004c, p25). It is interesting to compare the overall figures with Bouis's original projection in 1996 of a one-time spend of US$10 million over four years (Bouis, 1996a).

17. Interview, BMGF, 30 November 2005.

18. www.sustaintech.org/txtactivities.htm (27 July 2007).

19. www.rockfound.org/initiatives/index.shtml (23 July 2007).

20. www.rockfound.org/about_us/news/2007/0412rodin_google.shtml (18 March 2008).

21. Jane Wales, World Affairs Council of Northern California, quoted at: www.rockfound.org/about_us/news/2007/0412rodin_google.shtml (18 March 2008).

22. Jeffrey Sachs, quoted at: www.rockfound.org/about_us/news/2007/0408philanthropy.shtml (27 July 2007).

23. www.microsoft.com/presspass/exec/billg/bio.mspx (18 March 2008).

24. http://news.bbc.co.uk/1/hi/business/3428721.stm (8 August 2007).

25. www.nndb.com/people/533/000044401/ (18 March 2008).

26. http://online.wsj.com/public/article_print/SB116371407515425544.html (8 August 2007).

27. www.usnews.com/usnews/biztech/charities/articles/8stone.htm (18 March 2008). Stonesifer stepped down from her position as CEO of the BMGF in 2008.

28. http://foundationcenter.org/pnd/archives/19991116/003030.html (18 March 2008).

29. http://foundationcenter.org/pnd/archives/19991116/003030.html (18 March 2008).

30. www.gatesfoundation.org/AboutUs/OurValues/GatesLetter/ (18 March 2008).

31. Interview, HarvestPlus, 27 January 2006.

32. Interview, BMGF, 30 November 2005.

33. Interview, HarvestPlus, 27 January 2006.

34. www.cgiar.org/pdf/agm03stake_biofortificationcp_update.pdf (25 July 2007).

35. www.cgiar.org/pdf/agm03bus_harvestplus_bcp.pdf (25 July 2007).

36. www.cgiar.org/pdf/agm03bus_harvestplus_bcp.pdf (25 July 2007).

37. www.cgiar.org/exco/exco8/exco8_harvestplus_report.pdf (25 July 2007).

38. Interview, HarvestPlus, 27 January 2006.

39. www.cgiar.org/exco/exco8/exco8_harvestplus_report.pdf (25 July 2007).

40. The technical and policy uncertainties raised by alternative interpretations of

GxE (genotype by environment) interactions are discussed in Chapter 2 and again in later sections of this chapter.

41. Interview, HarvestPlus, 27 January 2006.
42. Interviews, HarvestPlus, 17, 26, 27 January 2006.
43. www.cgiar.org/pdf/agm03stake_biofortificationcp_update.pdf (25 July 2007).
44. HarvestPlus interview, 26 January 2006.
45. IR68144 refers to the iron-dense rice germplasm generated by IRRI and used as the experimental material for the nutritional study (Haas et al, 2005). MS13 refers to the name given to the variety, released in the Philippines, derived from IR68144 (Padolina et al, 2003). See Chapter 2 for further discussion of these events.
46. Suman Suhai of Gene Campaign, at: www.genecampaign.org/News/golden-rice.htm (18 March 2008).
47. www.gmwatch.org/profile1.asp?PrId=294 (18 March 2008).
48. www.genecampaign.org/News/golden-rice.htm (18 March 2008).
49. www.irri.org/about/irridir.staffbio.asp (18 March 2008).
50. 'CropDesign was founded in 1998 as a spin-off from VIB, the Flanders Institute for Biotechnology, and was financially backed until 2006 by a consortium of venture capital funds led by GIMV. In 2006, CropDesign was acquired by BASF Plant Science and is since then integrated in the international research network of BASF Plant Science. CropDesign currently employs about 70 people at its facilities in Gent, Belgium': www.cropdesign.com/general.php (1 June 2008).
51. Interview, IRRI, 15 June 2006.
52. www.irri.org/about/irridir/staffbio.asp (18 March 2008).
53. See Chapter 2 for a more detailed discussion of the departure of Gregorio and other members of the iron rice 'family', and its consequences.
54. 'Parminder Virk was senior research fellow, Punjab Agricultural University, India (1984–5); visiting fellow, University of Birmingham, UK (1985–6); and research fellow, University of Birmingham, UK (1987–99)': www.irri.org/about/irridir/staffbio.asp (1 June 2008).
55. www.irri.org/about/irridir/staffbio.asp (18 March 2008).
56. HarvestPlus interview, 17 January 2006.
57. Interview, World Bank, 24 January 2006 (original emphasis).
58. Interview, HarvestPlus, 17 January 2006.
59. Interview, nutritionist, 28 March 2006.
60. Interview, nutritionist, 28 March 2006.
61. Interview, IRRI, 1 June 2006. As discussed in Chapter 4, the issue of whether the bioefficacy trial did, in fact, provide such a definitive 'bioavailability number' is contested, since the claims of Haas et al are more modest in this respect (Haas et al, 2005).
62. Interview, IRRI scientist, 30 May 2006.
63. Interviews, IRRI scientists, 25 and 30 May 2006.
64. Interview, IRRI scientist, 25 May 2006.
65. IRRI's Strategic Plan for 2007–15 includes a target to increase consumption of brown or under-polished rice (IRRI, 2006, p32). However, the decision not to include it within the HarvestPlus programme left it vulnerable to the uncertainties of core funding.
66. Interview, IRRI, 29 May 2006.
67. Interview, IRRI scientist, 25 May 2006.
68. Interview, IRRI scientist, 30 May 2006.
69. Interview, IRRI scientist, 30 May 2006.

70. Interview, IRRI scientist, 18 December 2006.
71. These findings were from studies using soybean ferritin. However, similar results may be possible with rice ferritin contained in the plant leaves (but not in the grain). IRRI scientists have been exploring both options (Interviews, IRRI, 18 December 2006).
72. Interview, PhilRice, 5 June 2006.
73. Interview, IRRI scientist, 30 May 2006.
74. Interview, PhilRice scientist, 7 June 2006.
75. Interview, PhilRice scientist, 30 May 2006.
76. Interviews, NARS scientist, 16 January 2007; IRRI scientist, 7 December 2006.
77. Interview, NARS scientist, 16 January 2007.
78. Interview, nutritionist, 28th March 2006 (original emphasis).
79. Mackintosh et al (2008) have drawn attention to tensions and conflicts between the roles of 'broker' and 'gatekeeper'.
80. Personal communication, IRRI scientist, 23 May 2006.
81. Interview, IRRI scientist, 18 December 2006.
82. For example, recent 'end-user' research on biofortified orange sweet potato and maize has explored the use of 'choice experiments' in analysing the factors that influence choices between products with nutritional benefits (due to higher levels of pro-vitamin A) vis-à-vis other characteristics, such as appearance (colour) and texture. Findings were presented by J. V. Meenakshi, Impact and Policy Coordinator for HarvestPlus, at a STEPS Centre Seminar at the University of Sussex, UK, on 9 May 2008. See: www.steps-centre.org/events/stepsseminars.html (1 June 2008).
83. Interview, IRRI, 25 May 2006.
84. Florencio (2004) highlights the problematic nature of the term 'food-based', enabling its use by, for example, the industrial food fortification lobby, preferring the more precise term 'dietary-based' referring to changes in diet composition.
85. Interview, WHO, 14 March 2006.
86. While this research was in progress, exploratory discussions were taking place at IRRI regarding the possibility of partnering with NGOs and/or other development organizations to deliver biofortified varieties through targeted pro-poor interventions such as food-for-work programmes (IRRI interview, 2 June 2006).
87. Interview, WHO, 14 March, 2006.
88. www.goldenrice.org/Content1-Who/who_Matin.html (17 March 2008).
89. Interviews, HarvestPlus, 17 and 27 January 2006.
90. Donor interviews, 2 February and 25 January 2006.
91. www.harvestplus-china.org/english/swqh.htm (25 July 2007); www.harvestplus.org/pdfs/HPIndiaMOU.pdf (17 March 2008).
92. www.harvestplus-china.org/english/swqh.htm (25 July 2007).
93. Plenary question and answer session, Second Annual Meeting of the HarvestPlus-China programme, 18 September 2007.
94. A method for bioavailability testing *in vitro*, developed at Cornell University (Glahn et al, 2002).
95. Rice crop breakout group discussion, Second Annual Meeting of the HarvestPlus-China programme, 18 September 2007.
96. In this case the advice was to commission a bioavailability study using stable isotopes, rather than the more time-consuming bioefficacy study.
97. Rice crop breakout group discussion, Second Annual Meeting of the HarvestPlus-China programme, 18 September 2007.

98. www.goldenrice.org/Content2-How/how10_PVMRC.html (9 June 2007).

99. Peter Beyer (one of the Golden Rice co-inventors) conducted research on Vitamin E transformation; Michael Grusak (Baylor) on iron bioavailability; Dean Dellapenna (Michigan State) on nutritional genomics; Samuel Sun (Hong Kong) on lysine transformation. Meanwhile adaptive research on Golden Rice is ongoing at Cuu Long, PhilRice and IRRI (see Chapter 3 for further details).

100. Interview, NARS scientist, 16 January 2007.

101. www.goldenrice.org/Content2-How/how10_PVMRC.html (9 June 2007).

102. Interview, NARS scientist, 16 January 2007.

Chapter 5

1. For example, recent empirical findings from the Philippines have highlighted the endurance of seasonal protein-energy malnutrition, rather than micronutrient deficiencies, as the overriding nutritional concern (Frei and Becker, 2004). Furthermore, the broader nutritional benefits of promoting a wider diversity of landraces, in particular upland, pigmented rices, especially in their unmilled form, are also highlighted (Frei and Becker, 2004). An article in the journal *Food Policy*, co-authored by Timothy Johns of IPGRI and Pablo Eyzaguirre (2006), further develops the 'biotechnology versus biodiversity' theme, questioning assumptions that current biofortification programmes will complement existing approaches in an effective manner, arguing that 'focusing on staple cereals is unlikely to benefit the poor for economic and nutritional reasons alike, as it leads to even greater dietary simplification [which] can do more harm than good' (Johns and Eyzaguirre, 2006, p19).

2. Interviews, NARS scientists, 16 January 2007; IRRI scientist, 7 December 2006.

3. Interview, IRRI, 24 May 2006.

4. Interview, HarvestPlus, 17 January 2006.

5. Interview, Cornell University, 31 January 2008.

6. Interview, IRRI scientist, 30 May 2006.

7. IRRI interview, 29 May 2006.

8. Interview, NARS scientist, 16 January 2007.

9. Interview, HarvestPlus, 19 September 2007.

10. Donor interview, 25 January 2006.

11. www.copenhagenconsensus.com/Default.aspx?ID=788 (24 March 2008).

12. Derham at www.seedquest.com/News/2005/may/12362.htm (19 June 2007).

13. For example, see www.scidev.net/en/news/uk-to-streamline-health-aid-strategy.html (8 April 2008).

14. Interview, PhilRice, 30 May 2006.

15. Interview, IRRI scientist, 25 May 2006.

16. As discussed in Chapter 2, in the Philippines in the 1950s a key obstacle faced by a national programme to fortify rice with vitamin B was the resistance of rice millers, who saw it as a mechanism to monitor their income (Interview, FNRI, 21 June 2006).

Conclusion

1. www.harvestplus.org/content/worlds-top-economists-say-biofortification-one-top-five-solutions-global-challenges (16 May 2009).

2. www.harvestplus.org/content/harvestplus-receives-funding-new-research-phase (4 May 2009).

3.   www.harvestplus.org/content/worlds-top-economists-say-biofortification-
      one-top-five-solutions-global-challenges (16 May 2009).
4.   http://query.nytimes.com/gst/fullpage.html?res=9E0CE6DF1630F931A25753
      C1A96E9C8B63 (16 May 2009).
5.   Interview, nutritionist, 28 March 2008.
6.   This workshop (discussed in Chapter Two) was documented in a special issue of
      the *Food and Nutrition Bulletin* (vol 21, no 2, 2000).
7.   www.ifpri.org/pubs/books/oc58.asp (16 May 2009).

References

Al-Babili, S. and Beyer, P. (2005) 'Golden Rice – five years on the road – five years to go?', *Trends in Plant Science*, vol 10, pp565–73

Anderson, R. S., Levy, E. and Morrison, B. M. (1991) *Rice Science and Development Politics: Research Strategies and IRRI's Technologies Confront Asian Diversity (1950–1980)*, Clarendon Press, Oxford, UK

ANGAT-Laguna (2006) 'Rice-Biofortification (Iron-rich rice): A Community-based approach to specifically address Iron-Deficiency Anaemia (IDA). A joint project of ANGAT-Laguna, International Rice Research Institute (IRRI) and Philippine Rice Research Institute (PhilRice)', Project document available from the Office of the Vice-Governor, Laguna Province, Philippines

Apthorpe, R. (1996) 'Reading development policy and policy analysis: On framing, naming, numbering and coding', in R. Apthorpe and D. Gaspar (eds) *Arguing Development Policy: Frames and Discourses*, Frank Cass, London

Apthorpe, R. and Gasper, D. (1996) *Arguing Development Policy: Frames and Discourses*, Frank Cass, London

Ashby, J. A. (2009) 'Fostering farmer first methodological innovation: Organizational learning and change in international agricultural research', in I. Scoones and J. Thompson (eds) *Farmer First Revisited: Innovation for Agricultural Research and Development*, Practical Action, London, UK

Asia Rice Foundation (2004) *Rice in the Seven Arts*, Asia Rice Foundation, Los Baños, Philippines

Ayele, S., Chataway, J. and Wield, D. (2006) 'Commentary: Partnerships in African crop biotech', *Nature Biotechnology*, vol 24, no 6, June 2006, pp619–21

Balgos, C. C. A. (2005) 'Food and the Filipino: Feast and famine?', *Investigative Reporting Quarterly*, vol 1, pp2–7

Barker, C. and Green, A. (1996) 'Opening the debate on DALYs', *Health Policy and Planning*, vol 11, pp179–83

Barker, R. and Dawe, D. (2002) 'The transformation of the Asian rice economy and directions for future research: The need for increased productivity', in M. Sombilla, M. Hossain and B. Hardy (eds) *Developments in Asian Rice Economy. International Workshop on Medium- and Long-Term Prospects of Rice Supply in the 21st Century*, International Rice Research Institute, Los Baños, Philippines

BCP (2003) 'Update on Biofortification Challenge Program Activities, May 2003: First Project Advisory Committee Meeting and Selection of Program Director', CGIAR, Washington, DC

Behrman, J. R. (1995) 'Household behaviour and micronutrients: What we know and what we don't know', in H. E. Bouis (ed.) *Working Papers on Agricultural Strategies for Micronutrients*, International Food Policy Research Institute (IFPRI), Washington, DC

Behrman, J. R., Alderman, H. and Hoddinott, J. (2004) 'Hunger and malnutrition: Copenhagen Consensus Challenges Paper', Copenhagen Consensus, Copenhagen

Bello, W. (2004) *The Anti-Development State: The Political Economy of Permanent Crisis in the Philippines*, University of the Philippines, Diliman, and Focus on the Global South, Quezon City and Bangkok

Berber, J. (2003) 'Regional rice farmers' consultations', *Farm News and Views*, 2nd–3rd Quarter 2003, pp10–23

Bertaux, B. (1981) *Biography and Society: The Life History Approach in the Social Sciences*, Sage, London

Beyer, P., Al-Babili, S., Ye, X., Lucca, P., Schaub, P., Welsch, R. and Potrykus, I. (2002) 'Golden Rice: Introducing the beta-carotene biosynthesis pathway into rice endosperm by genetic engineering to defeat vitamin A deficiency', *Journal of Nutrition*, Symposium: Plant Breeding: A New Tool for Fighting Micronutrient Malnutrition, pp506s–510s

Bhaskaram, P. (2002) 'Micronutrient malnutrition, infection and immunity: An overview', *Nutrition Reviews*, vol 60, ppS40–S45

Biggs, S. D. and Clay, E. J. (1981) 'Sources of innovation in agricultural technology', *World Development*, vol 9, pp321–36

BIOTHAI (Thailand), CEDAC (Cambodia), DRCSC (India), GRAIN, MASIPAG (Philippines), PAN-Indonesia and UBINIG (Bangladesh) (2001) *Grains of Delusion: Golden Rice Seen from the Ground*, MASIPAG, Los Baños, Philippines

Bouis, H. (1995a) 'F.A.S. Public Interest Report: Breeding for nutrition', *Journal of the Federation of American Scientists*, vol 1, pp8–16

Bouis, H. (1995b) 'Enrichment of food staples through plant breeding: A new strategy for fighting micronutrient malnutrition', *SCN News*, vol 12, pp15–19

Bouis, H. (1996a) 'Enrichment of food staples through plant breeding: A new strategy for fighting micronutrient malnutrition', *Nutrition Reviews*, vol 54, 131–7

Bouis, H. (1996b) 'Plant breeding strategies for improving human mineral and vitamin nutrition', *Micronutrients and Agriculture*, no 1, pp1–2

Bouis, H. (2004) 'Hidden hunger: The role of nutrition, fortification and biofortification', World Food Prize International Symposium: From Asia to Africa: Rice, Biofortification and Enhanced Nutrition, Des Moines, IA

Bouis, H. (2006) 'HarvestPlus: Proposal 2008–11', HarvestPlus-China: Second Annual Meeting, Hangzhou, China

Bouis, H. E., Graham, R. D. and Welch, R. M. (1999) 'The CGIAR Micronutrients Project: Justification, history, objectives and summary of findings', Improving Human Nutrition through Agriculture: The Role of International Agricultural Research workshop, IRRI, Los Baños, Philippines

Boulton, L. and Lamont, J. (2007) 'Philanthropy "can eclipse G8" on poverty', *Financial Times*, 8 April

Brenner, C. (1993) *Technology and Developing Country Agriculture: The Impact of Economic Reform*, OECD, Paris

Brooks, S. (2005) 'Biotechnology and the politics of truth: From the Green Revolution to an "Evergreen Revolution"', *Sociologia Ruralis*, vol 45, pp360–79

Brooks, S. (2006) 'Biofortification and HarvestPlus: Some preliminary reflections', unpublished

Brooks, S. (2007) 'Sociological research on rice biofortification', Joint Divisional Seminar: Plant Breeding, Genetics and Biotechnology Division and Social Sciences Division, International Rice Research Institute, Los Baños, Philippines

Brooks, S. (2008) *Global Science, Public Goods? Tracing International Science Policy Processes in Rice Biofortification*, PhD thesis, University of Sussex, Brighton, UK

Brooks, S., Thompson, J., Odame, H., Kibaara, B., Nderitu, S., Karin, F. and Millstone, E. (2009) *Environmental Change and Maize Innovation in Kenya: exploring pathways in and out of maize*, STEPS Working Paper 36, Brighton: STEPS Centre

Brooks, S., Leach, M., Lucas, H. and Millstone, E. (2009a) *Silver bullets, Grand Challenges and the New Philanthropy*, STEPS Working Paper 24, Brighton: STEPS Centre

Brown, M. J., Ferruzzi, M. G., Nguyen, M. L., Cooper, D. A., Eldridge, A. L., Schwartz, S. J. and White, W. S. (2004) 'Carotenoid bioavailability is higher from salads ingested with full-fat than with fat-reduced salad dressings as measured with electrochemical detection', *American Journal of Clinical Nutrition*, vol 80, pp396–403

Bryce, J., Coitinho, D., Darnton-Hill, I., Pelletier, D. L. and Pinstrup-Andersen, P. (2008) 'Maternal and child undernutrition 4: Maternal and child undernutrition: Effective action at national level', *Lancet*, vol 371, pp510–26

Cakmak, I. (1996) 'Zinc deficiency as a critical constraint in plant and human nutrition in Turkey', *Micronutrients and Agriculture*, vol 1, pp13–14

Callon, M. (1986) 'Some elements of a sociology of translation: Domestication of the scallops and the fisherman of St. Brieuc Bay', in J. Law (ed.) *Power, Action and Belief: A New Sociology of Knowledge*, Routledge and Kegan Paul, London

Callon, M. (1991) 'Techno-economic networks and irreversibility', in J. Law (ed.) *A Sociology of Monsters: Essays on Power, Technology and Domination*, Routledge, London

Calloway, D. H. (1995) 'Human nutrition: Food and micronutrient relationships', in H. E. Bouis (ed.) *Working Papers on Agricultural Strategies for Micronutrients, No 1*, International Food Policy Research Institute, Washington, DC

Cantrell, R. P. and Reeves, T. G. (2002) 'The cereal of the world's poor takes centre stage', *Science*, vol 296, p53

Castillo, G. T. (2006) *Rice in Our Life: A Review of Philippine Studies*, Angelo King Institute, De La Salle University and Philippine Rice Research Institute, Manila, Philippines

CGIAR (2001) *CGIAR Reform Programme, 2001*, CGIAR Secretariat, World Bank, Washington, DC

Chambers, R., Pacey, A. and Thrupp, L. A. (eds) (1989) *Farmer First: Farmer Innovation and Agricultural Research*, Intermediate Technology Publications, London

Chandler, R. F. J. (1992) *An Adventure in Applied Science: A History of the International Rice Research Institute*, International Rice Research Institute, Manila, Philippines

Chataway, J., Smith, J. and Wield, D. (2007) 'Shaping scientific excellence in agricultural research', *International Journal of Biotechnology*, vol 9, pp172–87

CIAT and IFPRI (2002) 'Biofortified crops for improved human nutrition: A Challenge Program Proposal presented by CIAT and IFPRI to the CGIAR Science Council', International Centre for Tropical Agriculture and International Food Policy Research Institute, Washington, DC

CIMMYT (1995) 'Tackling zinc deficiency in Turkey', International Maize and Wheat Improvement Centre (CIMMYT), Mexico

Clay, E. J. and Schaffer, B. B. (1984a) 'Introduction: Room for manoeuvre: The premise of public policy', in E. J. Clay and B. B. Schaffer (eds) *Room for Manoeuvre: An Exploration of Public Policy Planning in Agricultural and Rural Development*, Fairleigh Dickinson University Press, Madison, NJ

Clay, E. J. and Schaffer, B. B. (1984b) 'Conclusion: Self awareness in policy practice', in E. J. Clay and B. B. Schaffer (eds) *Room for Manoeuvre: An Exploration of Public Policy Planning in Agricultural and Rural Development*, Fairleigh Dickinson University Press, Madison, NJ

Cohen, J. (1995) 'A Mexican-bred super maize', *Science*, vol 267, p825

Cohen, J. (2000) 'The Global Burden of Disease Study: A useful projection of future global health?', *Journal of Public Health Medicine*, vol 22, pp518–24

Combs, G. F., Welch, R. M., Duxbury, J. M., Uphoff, N. T. and Nesheim, M. C. (1996) 'Food-based approaches to preventing micronutrient malnutrition: An international research agenda', Cornell International Institute for Food, Agriculture and Development (CIIFAD), Cornell University, Ithaca, NY

Corpuz-Arocena, E. R., Abilgos-Ramos, R. G., Luciano, V. P., Justo, J. E., Manaois, R. V., Garcia, G. D., Escubio, S. S. P., Julaton, C. N., Alfonso, A. A., Padolina, T. F., Sebastian, L. S., Tabien, R. E., Cruz, R. T. and dela Cruz Jr, H. C. (2004) 'Breeding for micronutrient-dense rice in the Philippines', *PhilRice Technical Bulletin*, vol 8, p11

Cribb, J. (1995) 'Researchers sow seeds of nutrition', *Weekend Australian*, 11–12 February

Cullather, N. (2004) 'Miracles of modernization: The Green Revolution and the apotheosis of technology, *Diplomatic History*, vol 28, pp227–54

Dano, E. and Obanil, R. (2005) 'Attaining rice self sufficiency in the Philippines: Lessons from major rice producing countries', *State Intervention in the Rice Sector in Selected Countries: Implications for the Philippines*, Searice and Rice Watch and Action Network (R1), Quezon City, Philippines

Darnton-Hill, I. (1998) 'Overview: Rationale and elements of a successful food-fortification programme', *Food and Nutrition Bulletin*, vol 19, pp92–100

Darnton-Hill, I., Bloem, M. W., Benoist, B. D. and Brown, L. R. (2002) 'Micronutrient restoration and fortification: Communicating change, benefits and risks', *Asia Pacific Journal of Clinical Nutrition*, vol 11, ppS184–S196

Datinguinoo, V. M. (2005) 'Food and the Filipino: Feast and famine?' *Investigative Reporting Quarterly*, vol 1, pp84–91

Datta, K., Baisakh, N., Oliva, N., Torrizo, L., Abrigo, E., Tan, J., Rai, M., Rehana, S., Al-Babili, S., Beyer, P., Potrykus, I. and Datta, S. K. (2003) 'Bioengineered "golden" indica rice cultivars with beta-carotene metabolism in the endosperm with hygromycin and mannose selection systems', *Plant Biotechnology Journal*, vol 1, pp81–90

David, C. C. and Otsuka, K. (eds) (1993) *Modern Rice Technology and Income Distribution in Asia*, IRRI, Manila, Philippines

Davila-Hicks, P., Theil, E. C. and Lonnerdal, B. (2004) 'Iron in ferritin or in salts (ferrous sulphate) is equally bioavailable in nonanemic women', *American Journal of Clinical Nutrition*, vol 80, pp936–40

Dawe, D. D., Moya, P. F. and Casiwan, C. B. (2006) 'Executive summary', *Why Does the Philippines Import Rice? Meeting the Challenge of Trade Liberalization*, International Rice Research Institute and Philippine Rice Research Institute, Metro Manila and Nueva Ecija, Philippines

dela Cuadra, A. C. (2000) 'The Philippine micronutrient supplementation programme', *Food and Nutrition Bulletin*, vol 21, pp512–14

Delisle, D. (2003) 'Food diversification strategies are neglected in spite of their potential effectiveness: Why is it so and what can be done?' 2nd International Workshop, Food-based Approaches for a Healthy Nutrition, Ouagadougou

Del Mundo, A. M. (2003) 'Closing remarks', *Proceedings: Feedback Seminar to Rice Feeding Study Participants*, Institute of Nutrition and Food, College of Human Ecology, University of the Philippines Los Baños, Pasay City, Philippines

Department of Health (2007) 'Health is global: Proposals for a UK government-wide strategy – A report from the UK's Chief Medical Advisor Sir Liam Donaldson', Department of Health, London

Economist (2006) 'Rockefeller revolutionary: Judith Rodin is shaking up one of the world's venerable charitable foundations', *Foundation News*, Rockefeller Foundation, New York

Edwards, M. (2008) *Just Another Emperor? The Myths and Realities of Philanthrocapitalism*, Demos and The Young Foundation, New York

Eicher, C. K. and Rukuni, M. (2003) 'Thematic working paper: The CGIAR in Africa: Past present and future, in World Bank Operations Evaluation Department (ed) *The CGIAR at 31: An Independent Meta-evaluation of the Consultative Group on International Agricultural Research*, World Bank, Washington, DC

Evenson, R. E., Herdt, R. W. and Hossain, M. (1996) *Rice Research in Asia: Progress and Priorities*, CAB International, Wallingford, UK

FAO (2003) 'The International Year of Rice 2004: Concept paper', International Year of Rice Secrerariat, Food and Agriculture Organization of the United Nations, Rome

Fiedler, J. L., Dado, D. R., Maglalang, H., Juban, N., Capistrano, M. and Magpantay, M. V. (2000) 'Cost analysis as a vitamin A programme design and evaluation tool: A case study of the Philippines', *Social Science Medical Journal*, vol 51, pp223–42

Fischer, F. (1998) 'Beyond empiricism: Policy inquiry in a postpositivist perspective', *Policy Studies Journal*, vol 26, pp129–46

Fischer, F. (2003) *Reframing Public Policy: Discursive Politics and Deliberative Practices*, Oxford University Press, Oxford

Fitzhugh, H. and Brader, L. (2002) 'Core funding for systemwide and ecoregional programmes: Draft submitted to the CGIAR on 31st March 2002', CGIAR, Washington, DC

Florencio, C. A. (2000) 'Nutrition problems and causes: A study of two cases', in K. Krishnaswamy (ed.) *Nutrition Research: Current Scnerario and Future Trends*, IBH Publishing Co., New Delhi and Oxford

Florencio, C. A. (2004) *Nutrition in the Philippines: The Past for its Template, Red for its Colour*, University of the Philippines Press, Diliman, Quezon City, Philippines

FNRI (2003) *Philippine Nutrition: Facts and Figures 2003*, Food and Nutrition Research Institute, Department of Science and Technology (DOST), Bicutan, Taguig City, Philippines

Foucault, M. (1976) 'Right of death and power over life', *The History of Sexuality*, Picador, London

Frankel, F. R. (1971) *India's Green Revolution: Economic Gains and Political Costs*, Princeton University Press, Princeton

Frankel, F. R. (1974) 'The politics of the Green Revolution: Shifting patterns of peasant participation in India and Pakistan', in T. T. Poleman and D. K. Freebairn (eds) *Food, Population and Employment: The Impact of the Green Revolution*, Praeger Publishers, Westport, CT

Frei, M. and Becker, K. (2004) 'Agro-biodiversity in subsistence-oriented farming systems in a Philippine upland region: Nutritional considerations', *Biodiversity and Conservation*, vol 13, pp1591–610

Fujisaka, S. (1994) 'Will farmer participatory research survive in the agricultural research centres?', in I. Scoones and J. Thompson (eds) *Beyond Farmer First: Rural People's Knowledge, Agricultural Research and Extension Practice*, Intermediate Technology Publications, London

GAIN (2005) *GAIN Annual Report 2004–2005*, Global Alliance for Improved Nutrition, Geneva

Gasper, D. (1996) 'Analysing policy arguments', in R. Apthorpe and D. Gaspar (eds) *Arguing Development Policy: Frames and Discourses*, Frank Cass, London

Gates Foundation (1999) *Annual Report*, Gates Foundation, Seattle

Gates Foundation (2000) *Annual Report*, Gates Foundation, Seattle

Gates Foundation (2001) *Annual Report*, Gates Foundation, Seattle

Gates Foundation (2002) *Annual Report*, Gates Foundation, Seattle

Gates Foundation (2005) 'The face of change: 22 stories from 2005', *Annual Report*, Gates Foundation, Seattle

Gates Foundation (2006) *Annual Report*, Gates Foundation, Seattle

Gieryn, T. F. (1999) *Cultural Boundaries of Science: Credibility on the Line*, University of Chicago Press, Chicago

Gillespie, S. and Mason, J. (1991) 'Nutrition-relevant actions: Some experiences from the eighties and lessons for the nineties', *ACC/SCN State-of-the-Art Series: Nutrition Policy Discussion Paper No. 10*, United Nations System Standing Committee on Nutrition (SCN), Geneva

Gillespie, S., McLachlan, M. and Shrimpton, R. (2004) *Combating Nutrition: Time to Act*, World Bank, Washington, DC

Glaeser, B. (1987) 'Agriculture between the Green Revolution and eco-development: Which way to go?', in B. Glaeser (ed.) *The Green Revolution Revisited: Critique and Alternatives*, Allen and Unwin, London

Glahn, R. P., Cheng, Z., Welch, R. M. and Gregorio, G. B. (2002) 'Comparison of iron bioavailability from 15 rice genotypes: Studies using an in vitro/Caco-2 cell culture model', *Journal of Agricultural and Food Chemistry*, vol 50, pp3586–91

Godin, B. (2006) 'The linear model of innovation: The historical construction of an analytical framework', *Science, Technology, and Human Values*, vol 31, pp639–67

Graham, R. D. (2002) *A Proposal for IRRI to Establish a Grain Quality and Nutrition Research Centre. Discussion Paper No. 44*, International Rice Research Institute, Manila, Philippines

Graham, R. D. and Welch, R. M. (1996) 'Breeding for staple food crops with high micronutrient density', in H. E. Bouis (ed.) *Working Papers on Agricultural Strategies for Micronutrients*, International Food Policy Research Institute (IFPRI), Washington, DC

Graham, R. D., Senadhira, D., Beebe, S., Iglesias, C. and Monasterio, I. (1999) 'Breeding for micronutrient density in edible portions of staple food crops: Conventional approaches', *Field Crops Research*, vol 60, pp57–80

Greenland, D. G. (1997) 'International agricultural research and the CGIAR system – Past present and future', *Journal of International Development*, vol 9, pp459–82

Gregorio, G. B. (2006) 'Living life to the fullest', in Q. N. Lee-Chua and L. S. Sebastian (eds) *In Love with Science: Outstanding Young Filipino Scientists Tell Their Stories*, Anvil Publishing Inc., Manila, Philippines

Gregorio, G. B. and Haas, J. D. (2005) 'Nutritional revolution in rice: A new scientific challenge', *Report 2005: Nestlé Foundation for the Study of Problems of Nutrition in the World*, Nestlé Foundation, Lausanne, Switzerland

Gregorio, G. B., Senadhira, D., Htut, H. and Graham, R. D. (2000) 'Breeding for trace mineral density in rice', *Food and Nutrition Bulletin*, vol 21, pp382–6

Gregorio, G. B., Sison, C. B., Mendoza, R. D., Adorada, D. L., Francisco, A. S., Escote, M. M. and Macabenta, J. T. (2003) 'Final report on production, milling and cooking trials to assess the Fe and Zn content of IR68144 and PSBRc28', Plant Breeding, Genetics and Biochemistry Division, International Rice Research Institute, Los Baños, Philippines

Griffin, K. (1979) *The Political Economy of Agrarian Change: An Essay on the Green Revolution*, Macmillan Press, New York

Gupta, A. and Ferguson, J. (1997) 'Discipline and practice: "The Field" as site, method and location in anthropology', in A. Gupta and J. Ferguson (eds) *Anthropological Locations: Boundaries and Grounds of a Field Science*, University of California Press, Berkeley, CA

Gwatkin, D. R. (1997) 'Correspondence', *Lancet*, vol 350, p141

Haas, J. D., Beard, J. L., Murray-Kolb, L. E., Del Mundo, A. M. Felix, A. and Gregorio, G. B. (2005) 'Iron-biofortified rice improves the iron stores of non-anaemic Filipino women', *Community and International Nutrition*, pp2823–30

Haas, J. D., Del Mundo, A. M. and Beard, J. L. (2000) 'A human feeding trial of iron-enhanced rice', *Food and Nutrition Bulletin*, vol 21, pp440–4

Hacking, I. (1990) *The Taming of Chance (Ideas in Context)*, Cambridge University Press, Cambridge

Haddad, L. (2000) 'A conceptual framework for assessing agriculture-nutrition linkages', *Food and Nutrition Bulletin*, vol 21, pp367–73

Hall, A. (2007) 'The origins and implications of using innovation systems perspectives in the design and implementation of agricultural research projects: Some personal observations', *Working Paper Series*, UNU-MERIT, Maastricht, Netherlands

HarvestPlus (2004a) *Breeding Crops for Better Nutrition: Harnessing Agricultural Technology to Improve Micronutrient Deficiencies*, International Food Policy Research Institute, Washington, DC

HarvestPlus (2004b) *Report on Activities January, 2003 through March, 2004*, submitted to World Bank and the Executive Council and Science Council of the CGIAR, International Food Policy Research Institute, Washington, DC

HarvestPlus (2004c) *HarvestPlus. Progress Report: May 2004–April 2005*, CGIAR, International Food Policy Research Institute, Washington, DC

Hawkes, C. and Ruel, M. T. (2006a) 'Agriculture and nutrition linkages: Old lessons and new paradigms. Brief 4 of 16', in C. Hawkes and M. T. Ruel (eds) *Understanding the Links between Agriculture and Health: 2020 Vision/Focus 13*, IFPRI, Washington, DC

Hawkes, C. and Ruel, M. T. (2006b) *Understanding the Links between Agriculture and Health, 2020 Vision/Focus 13*, International Food Policy Research Institute, Washington, DC

Herdt, R. (1996) 'Establishing the Rockefeller Foundation's priorities for rice biotechnology research in 1995 and beyond', *Rice Genetics III. Proceedings of the Third International Rice Genetics Symposium*, International Rice Research Institute, Manila, Philippines

Herdt, R., Toenniessen, G. and O'Toole, J. (2005) 'Plant biotechnology for developing countries, *Handbook of Agricultural Economics Volume III*, Elsevier, Amsterdam

Hilchey, T. (1995) 'Scientist seeks way to enrich basic crops in poor lands', *New York Times*, 29 January 1995

Hindmarsh, S. and Hindmarsh, R. (2002) 'Laying the molecular foundations of GM rice across Asia', *Resource Book Volume 1*, Malaysia Pesticide Action Network, Penang, Malaysia

Hoa, T. T. C., Al-Babili, S., Schaub, P., Potrykus, I. and Beyer, P. (2003) 'Golden Indica and Japonica rice lines amenable to deregulation', *Plant Physiology*, vol 133, pp161–9

Hornedo, F. H. (2004) 'Overview of rice in Philippine culture', *Rice in the Seven Arts*, Asia Rice Foundation, Los Baños, Philippines

Horton, R. (2008) 'Comment: Maternal and child undernutrition: an urgent opportunity', *Lancet*, vol 371, p179

Horton, S. and Ross, J. (2003) 'The economics of iron deficiency', *Food Policy*, vol 28, pp51–75

IFPRI (1995) 'Mineral-packed crops coming', *International Agricultural Development*, March–April 1995, p21

IFPRI (2003) 'Biofortification gets a boost from the Gates Foundation', *IFPRI Forum*, December 2003, p6

IFPRI (2005) *Proceedings of an International Dialogue on Pro-Poor Public-Private Partnerships for Food and Agriculture*, International Food Policy Research Institute, Washington, DC

IRRI (1996) *Annual Report 1995–6: Listening to the Farmers*, IRRI, Manila, Philippines

IRRI (1998) *Annual Report 1997–8: Biodiversity – Maintaining a Balance*, IRRI, Manila, Philippines

IRRI (1999) *Annual Report 1998–9: Rice: Hunger or Hope?*, IRRI, Manila, Philippines

IRRI (2000a) 'The Sisters of Nutrition', April 2000, IRRI, Manila, Philippines

IRRI (2000b) *Annual Report 1999-2000: The Rewards of Rice Research*, IRRI, Manila, Philippines

IRRI (2006) *Bringing Hope, Improving Lives: Strategic Plan 2007–2015*, IRRI, Manila, Philippines

IVACG (2003) 'Improving the vitamin A status of populations', XXI International Vitamin A Consultative Group Meeting, Marrakesh, Morocco

Jasanoff, S. (2005) '"Let them eat cake": GM foods and the democratic imagination', in M. Leach, I. Scoones and B. Wynne (eds) *Science and Citizens: Globalization and the Challenge of Engagement*, Zed Books, London

Johns, T. and Eyzaguirre, P. B. (2006) 'Biofortification, biodiversity and diet: A search for complementary applications against poverty and malnutrition', *Food Policy*, vol 32, pp1–24

Johnson-Welch, C., MacQuarrie, K. and Bunch, S. (2005) *A Leadership Strategy for Reducing Hunger and Malnutrition in Africa: The Agriculture/Nutrition Advantage*, International Centre for Research on Women, Washington, DC

Juma, C. and Serageldin, I. (2007) *Freedom to Innovate: Biotechnology and Africa's Development. A Report of the High-Level African Panel on Modern Biotechnology*, African Union (AU) and NEPAD, Addis Ababa, Ethiopia and Pretoria, South Africa

Kennedy, E. and Bouis, H. E. (1993) *Linkages between Agriculture and Nutrition: Implications for Policy and Research*, IFPRI, Washington, DC

Khush, G. S. (1998) 'Dedication', *Rice Genetics Newsletter*, vol 15, p7

King, J. C. (2002) 'Evaluating the impact of plant biofortification on human nutrition', *Journal of Nutrition*, vol 132, pp511s–513s

Knight, J. (2003) 'Crop improvement: A dying breed', *Nature*, vol 421, pp568–70

Knorr-Cetina, K. (1999) *Epistemic Cultures: How the Sciences Make Knowledge*, Harvard University Press, Cambridge, MA

Kryder, R. D., Kowalski, S. P. and Krattiger, A. F. (2000) *The Intellectual and Technical Property Components of Pro-Vitamin A Rice (Golden Rice): A Preliminary Freedom-To-Operate Review*, International Service for the Acquisition of Agri-biotech Applications, Ithaca, NY

Laguna, R. J. P. (2002) 'Marketing *Mestizo*', *PhilRice Newsletter*, vol 15, p8

Latour, B. (1987) *Science in Action*, Harvard University Press, Cambridge, MA

Latour, B. (2003) *Reassembling the Social: An Introduction to Actor-Network Theory*, Oxford University Press, Oxford, UK

Leach, M. and Scoones, I. (2006) *The Slow Race: Making Technology Work for the Poor*, Demos, London, UK

Lehmann, V. (2001) 'The Rockefeller Foundation', *Biotechnology and Development Monitor*, no 44/45, p17

Lipton, M. and de Kadt, E. (1988) *Agriculture-Health Linkages*, WHO, Geneva

Lipton, M. and Longhurst, R. (1989) *New Seeds and Poor People*, Unwin Hyman, London

Low, J., Walker, T. and Hijmans, R. (2001) 'The potential impact of orange-fleshed sweet potatoes on vitamin A intake in Sub-Saharan Africa', Regional workshop on food based approaches to human nutritional deficiencies. The VITAA Project, vitamin A and orange-fleshed sweet potatoes on vitamin A intake in Sub-Saharan Africa, VITAA, Nairobi, Kenya

Mackintosh, M., Chataway, J. and Wuyts, M. (2008) 'Promoting innovation, productivity and industrial growth and reducing poverty: Bridging the policy gap', *European Journal of Development Research*, vol 19, p1012

Maitland, A. (1995) 'Minerals enlisted to fight hunger', *Financial Times*, 24 January

Marcus, G. E. (1995) 'Ethnography in/of the World System: The emergence of multi-sited ethnography', *Annual Review of Anthropology*, vol 24, pp95–117

Marcus, G. E. (1998) 'The uses of complicity in the changing *mise-en-scène* of anthropological fieldwork', *Ethnography through Thick and Thin*, Princeton University Press, Princeton, NJ

Mason, J. B., Lotfi, M., Dalmiya, N., Sethuraman, K. and Deitchler, M. (2001) *The Micronutrient Report: Current Progress and Trends in the Control of Vitamin A, Iodine and Iron Deficiencies*, Micronutrient Initiative/International Development Research Centre, Ottawa, Canada

Morris, S., Cogill, B. and Uauy, R. (2008) 'Effective international action against undernutrition: Why has it proven so difficult and what can be done to accelerate progress?' *Lancet*, vol 371, pp608–21

Murray, C. J. L. and Acharya, A. K. (1997) 'Understanding DALYs', *Journal of Health Economics*, vol 16, pp703–30

Murray, C. J. L. and Lopez, A. D. (eds) (1996) *The Global Burden of Disease: A Comprehensive Assessment of Mortality and Disability from Diseases, Injuries and Risk Factors in 1990 and Projected in 2020*, Harvard University Press, Cambridge, MA

Murray, C. J. L. and Lopez, A. D. (1997) 'Global mortality, disability, and the contribution of risk factors: Global Burden of Disease Study', *Lancet*, vol 349, pp1436–42

Nash, M. (2000) 'Grains of Hope', *Time*, 31 July, pp38–46

Nestle, M. (2001) 'Article: Genetically engineered "Golden Rice" is unlikely to overcome vitamin A deficiency; Response by Ingo Potrykus', *Journal of the American Diabetic Association*, vol 101, pp289–90

News Republic (1995) 'Amber waves of grain may get new potency', *News Republic*, 29 January

News-Sun (1995) 'Amber waves of grain may get new potency', *News-Sun*, 29 January

Normile, D. (1999) 'Rice biotechnology: Rockerfeller to end network after 15 years of success', *Science*, vol 286, pp1468–9

Norton, G. W. and Francisco, S. R. (2006) 'Seed system, biotechnology and nutrition', in A. M. Balisacan, L. S. Sebastian and Associates (eds) *Securing Rice, Reducing Poverty*, SEARCA (Southeast Centre for Graduate Study and Research in Agriculture), Los Baños, Philippines

Oasa, E. K. (1987) 'The political economy of international agricultural research: A review of the CGIAR's response to criticism of the Green Revolution', in B. Glaeser (ed.) *The Green Revolution Revisited: Critique and Alternatives*, Allen and Unwin, London

Obanil, R. and Dano, E. (2005) 'Attaining rice self sufficiency in the Philippines: Lessons from major rice producing countries', in SEARICE/R1 (ed.) *State Intervention in the Rice Sector in Selected Countries: Implications for the Philippines*, Southeast Asia Regional Initiative for Community Empowerment (SEARICE) and Rice Watch and Action Network (R1), Quezon City, Philippines

Observer-Dispatch (1995) 'Amber waves of grain may get new potency', *Observer-Dispatch*, 29 January

Okie, S. (2006) 'Global health: The Gates–Buffett effect', *New England Medical Journal*, vol 355, pp1084–8

O'Toole, J. C., Toenniessen, G. H., Murashige, T., Harris, R. R. and Herdt, R. W. (2001) 'The Rockefeller Foundation's International Programme on Rice Biotechnology', in G. S. Khush, D. S. Brar and B. Hardy (eds) *Rice Genetics IV. Proceedings of the Fourth International Rice Genetics Symposium*, International Rice Research Institute, Los Baños, Philippines

Paarberg, R. L. (2009) *Starved for Science: How Biotechnology is being Kept out of Africa*, Harvard University Press, Cambridge, MA

Pacey, A. and Payne, P. (eds) (1981) *Agricultural Development and Nutrition*, FAO/UNICEF, Rome and New York

Padolina, T., Corpuz, E. R., Abilgos, R. G., Manaois, R. V., Escubio, S. S. P., Garcia, G. d., Luciano, V. P., Sebastian, L. S., Tabien, R. E., Leon, J. C. D., Gregorio, G. B. and Sison, C. B. (2003) *MS13, a Conventionally Bred Rice Line with Enhanced Micronutrient Content*, Philippine Rice Research Institute, Nueva Ecija, Philippines

Padolina, W. G. (2003) 'Looking beyond the rice feeding trial: Perspectives on the research implications', *Proceedings: Feedback Seminar to Rice Feeding Study Participants*, Institute of Nutrition and Food, College of Human Ecology, University of the Philippines Los Baños, Pasay City, Philippines

Paine, J. A., Shipton, C. A., Chaggar, S., Howells, R. M., Kennedy, M. J., Vernon, G., Wright, S. Y., Hinchcliffe, E., Adams, J. L., Silverstone, A. L. and Drake, R. (2005) 'Improving the nutritional value of Golden Rice through increased pro-vitamin A content', *Nature Biotechnology*, vol 23, pp482–7

PATH (2005) *The Research Behind the UltraRice Technology*, PATH, Seattle

PATH (2006) *Technology Solutions for Global Health: UltraRice*, PATH, Seattle

Pearse, A. (1980) *Seeds of Plenty, Seeds of Want: Social and Economic Implications of the Green Revolution*, Clarendon Press, Oxford, UK

Pelletier, D. L. (1995) 'The food-first bias and nutrition policy: Lessons from Ethiopia', *Food Policy*, vol 20, pp279–98

Perkins, J. H. (1997) *Geopolitics and the Green Revolution: Wheat, Genes and the Cold War*, Oxford University Press, Oxford, UK

Perlas, N. and Vellvé, R. (1997) *Oryza Nirvana? An NGO review of the International Rice Research Institute in Southeast Asia*, Searice, Manila

Pfeiffer, W. H. (2006) 'HarvestPlus: Breeding', HarvestPlus-China: Second Annual Meeting, Hangzhou, China

PhilRice (2002) 'PGMA transfers PhilRice to Palace', *PhilRice Newsletter*, vol 15, p20

PhilRice (2003) 'NE farmers see no risk in Golden Rice', *PhilRice Newsletter*, vol 16, p21

PhilRice (2006) *Hybrid Rice*, Philippine Rice Research Institute, Nueva Ecija, Philippines

Pinstrup-Andersen, P. (1981) *Nutritional Consequences of Agricultural Projects: Conceptual Relationships and Assessment Approaches*, World Bank, Washington, DC

Pollan, M. (2001) 'The great yellow hype', *New York Times Magazine*, 4 March

Potrykus, I. (2001) 'Golden Rice and beyond', *Plant Physiology*, vol 125, pp1157–61

Prah Ruger, J. (2005) 'The changing role of the World Bank in global health', *American Journal of Public Health*, vol 95, pp60–70

Pretty, J. N. and Chambers, R. (1994) 'Towards a learning paradigm: New professionalism and institutions for a sustainable agriculture', in I. Scoones and J. Thompson (eds) *Beyond Farmer First: Rural People's Knowledge, Agricultural Research and Extension Practice*, Intermediate Technology Publications, London

Richards, P., de Bruin-Hoekzema, M., Hughes, S. G., Kudadjie-Freeman, C., Kwame Offei, S., Struik, P. C. and Afio Zannou, A. (2009) 'Seed systems for African food security: Linking molecular genetic analysis and cultivator knowledge in West Africa', *International Journal of Technology Management*, vol 45, nos 1–2

Rijsberman, F. (2002) 'CGIAR Challenge Program on Water and Food: Business Plan', Discussion document for CP Water and Food Consortium Meeting on 13–14 June 2002, Columbo

Rogers, E. M. (2003) *Diffusion of Innovations*, Free Press, New York

Saith, A. (2006) 'From universal values to Millennium Development Goals: Lost in translation', *Development and Change*, vol 37, pp1167–99

Schaffer, B. (1984) 'Towards responsibility: Public policy in concept and practice', in E. J. Clay and B. B. Schaffer (eds) *Room for Manoevre: An Exploration of Public Policy Planning in Agricultural and Rural Development*, Fairlcigh Dickinson University Press, Madison, NJ

Schön, D. A. and Rein, M. (1994) *Frame Reflection: Towards the Resolution of Intractable Policy Controversies*, Basic Books/Harper Collins, London

Science Council (2006) *Summary Report on System Priorities for CGIAR Research 2005–2015*, Science Council Secretariat, FAO, Rome

Science Council and CGIAR Secretariat (2004) *Synthesis of Lessons Learned from Initial Implementation of the CGIAR Pilot Challenge Programs*, Science Council and CGIAR Secretariat, Washington, DC

Science Council Secretariat (2005) *Center and Challenge Program Medium Term Plans 2006-2008: Overview of the SC Commentary*, CGIAR, Washington, DC

SCN (2004a) *5th Report on the World Nutrition Situation: Nutrition for Improved Development Outcomes*, United Nations System Standing Committee on Nutrition (SCN), Geneva

SCN (2004b) 'Nutrition and the Millennium Development Goals', *SCN News*, vol 28, pp11–14

Sen, A. K. (1981) *Poverty and Famines*, Oxford University Press, Oxford

Seshia, S. and Scoones, I. (2003) 'Tracing policy connections: The politics of knowledge in the Green Revolution and biotechnology eras in India', *IDS Working Paper 188*, Institute of Development Studies, Brighton, UK

Shore, C. and Wright, S. (1997) 'Introduction: Policy: A new field of anthropology', in Wright S. and Shore C. (eds) *Anthropology of Policy: Critical Perspectives on Governance and Power*, Routledge, London

Simmonds, N. W. (1991) 'Selection for local adaptation in a plant breeding programme', *Theoretical and Applied Genetics*, vol 82, pp363–7

Slingerland, M., Koning, N., Merx, D. and Nout, R. (2003) 'Food-based approaches for reducing micronutrient malnutrition', *North–South Discussion Paper No 2*. North-South Centre, Wageningen, Netherlands

Solon, F. S. (2000) 'Food fortification in the Philippines: Policies, programmes, issues and prospects', *Food and Nutrition Bulletin*, vol 21, pp515–20

Solon, F. S., Tomas, M. D., Latham, M. C. and Popkin, B. M. (1979) 'An evaluation of strategies to control vitamin A deficiency in the Philippines', *American Journal of Clinical Nutrition*, 32, pp1445–53

Sommer, A. and Davidson, F. R. (2002) 'Assessment and control of vitamin A deficiency: The Annecy Accords', *Journal of Nutrition*, Proceedings of the XX International Vitamin A Consultative Group Meeting, pp2845s–2850s

Sommer, A., Katz, J. and Tarwotfo, I. (1984) 'Increased risk of respiratory disease and diarrhea in children with preexisting mild vitamin A deficiency', *American Journal of Clinical Nutrition*, vol 40, pp1090–5

Southern Tagalog Herald (2006) 'Kontra Anemia', *Southern Tagalog Herald*, vol XI, no 40, pp21–27

Spitz, P. (1987) 'The Green Revolution re-examined in India', in B. Glaeser (ed.) *The Green Revolution Revisited: Critique and Alternatives*, Allen and Unwin, London

Stein, A. J., Meenakshi, J. V., Qaim, M., Nestel, P., Sachdev, H. P. S. and Bhutta, Z. A. (2005) *Analysing the Health Benefits of Biofortified Staple Crops by Means of the Disability-Adjusted Life Years Approach: A Handbook Focusing on Iron, Zinc and Vitamin A*, International Centre for Tropical Agriculture and International Food Policy Research Institute, Washington, DC

Stein, A. J., Qaim, M., Meenakshi, J. V., Nestel, P., Sachdev, H. P. S. and Bhutta, Z. A. (2006) 'Potential impacts of iron biofortification in India', *Discussion Paper No 04/2006*, University of Hohenheim, Hohenheim, Germany

Stoop, W. A. and Kassam, A. H. (2006) 'The "System of Rice Intensification (SRI)": Implications for agronomic research', *Tropical Agriculture Association Newsletter*, no 26, pp22–4

Strathern, M. (1995) 'Forward', *Shifting Contexts: Transformations in Anthropological Knowledge*, Routledge, London

Sumberg, J. (2005) 'Systems of innovation theory and the changing architecture of agricultural research in Africa', *Food Policy*, vol 30, pp21–41

Tabor, S. R. (1995) *Agricultural Research in an Era of Adjustment: Policies, Institutions, and Progress*, Economic Development Institute of the World Bank/International Service for National Agricultural Research, Washington, DC

Tang, G., Qin, J., Dolnikowski, G. G., Russel, R. M., Grusak, M. A. (2009) 'Golden Rice is an effective source of vitamin A', *American Journal of Clinical Nutrition*, vol 89, pp1776–83

Taverne, R. (2007) 'The real GM food scandal', *Prospect*, vol 140, pp24–7

Times-Union (1995) 'Enriched crops, soil may improve Third-World diets', *Times-Union*, 29 January

Toenniessen, G. (2000) *Vitamin A Deficiency and Golden Rice: The Role of the Rockefeller Foundation*, Rockefeller Foundation, New York

Tolentino, V. B. J. (2002) 'Governance constraints to sustainable rice productivity in the Philippines', in M. Sombilla, M. Hossain and B. Hardy (eds) *Developments in Asian Rice Economy. International Workshop on Medium- and Long-Term Prospects of Rice Supply in the 21st Century*, International Rice Research Institute, Los Baños, Philippines

Tontisirin, K., Attig, G. A. and Winichagoon, P. (1995) 'An eight-stage process for national nutrition development', *Food and Nutrition Bulletin*, vol 16, no 1, pp8–16

Tsing, A. (2002) 'Conclusion: The global situation', in J. X. Inda and R. Rosaldo (eds) *The Anthropology of Globalization: A Reader*, Blackwell, London

Uphoff, N. (2004) 'IRRI "Grain of Truth" exchange on the SRI controversy: "System of Rice Intensification responds to 21st Century needs". Response by T. Sinclair "Agronomic UFOs waste valuable scientific resources"', *Rice Today*, vol 3, pp42–3

Van Lieshout, M., West, C. E., Permaesih, M. D., Wang, Y., Xu, X., Breeman, R. B. V., Creemers, A. F. L., Verhoeven, M. A. and Lugtenburg, J. (2002) 'Bioefficacy of beta-carotene dissolved on oil studied in children in Indonesia', *American Journal of Clinical Nutrition*, vol 73, no 5, pp949–58

Van Roozendaal, G. (1996) 'Enhancing the nutritional qualities of crops: A second Green Revolution?' *Biotechnology and Development Monitor*, vol 29, pp12–15

von Braun, J., Fan, S., Meinzen-Dick, R., Rosegrant, M. W. and Nin Pratt, A. (2008) 'Agricultural Research for Food Security, Poverty Reduction and the Environment: What to Expect from Scaling Up CGIAR Investments and 'Best Bet' Programmes', IFPRI Issue Brief no 53, International Food Policy Research Institute, Washington, DC

Welch, R. M. (1996) 'Viewpoint: The optimal breeding strategy is to increase the density of promoter compounds and micronutrient mineral in seeds; Caution should be used in reducing anti-nutrients in staple food crops', *Micronutrients and Agriculture*, vol 1, pp 20–2

Welch, R. M. and Graham, R. D. (2002) 'Breeding crops for enhanced micronutrient content', *Plant and Soil*, vol 245, pp205–214

Welch, R. M. and Graham, R. D. (2004) 'Breeding for micronutrients in staple crops from a human nutrition perspective', *Journal of Experimental Botany*, vol 55, pp353–64

Welch, R. M., House, W. A., Beebe, S., Senadhira, D., Gregorio, G. B. and Cheng, Z. (2000) 'Testing iron and zinc bioavailability in genetically enriched beans (*Phaseolus vulgaris* L.) and rice (*Oryza sativa* L.) in a rat model', *Food and Nutrition Bulletin*, vol 21, pp428–33

West, C. E., Eilander, A. and Van Lieshout, M. (2002) 'Consequences of revised estimates of carotenoid bioefficacy for dietary control of vitamin A deficiency in developing countries', *American Society for the Nutritional Sciences Journal of Nutrition*, vol 132, pp2920S–2926S

World Bank (1993) *World Development Report 1993: Investing in Health*, Oxford University Press, New York

World Bank (2006) *Re-positioning Nutrition as Central to Development: A Strategy for Large-scale Action*, World Bank, Washington, DC

World Bank (2007) 'Meeting growing demand for agriculture through innovations in science and technology', *World Bank Development Report 2008: Policy Brief*, World Bank, Washington, DC

Wright, S. (1995) 'Anthropology: Still the uncomfortable discipline?', in A. S. Ahmed and C. Shore (eds) *The Future of Anthropology: Its Relevance to the Contemporary World*, Athlone, London

Ye, X., Al-Babili, S., Kloti, A., Zhang, J., Lucca, P., Beyer, P. and Potrykus, I. (2000) 'Engineering the provitamin A (beta-carotene) biosynthetic pathway into (caretenoid-free) rice endosperm', *Science*, vol 287, pp303–5

Yudelman, M., Blake, R. O., Bell, D. E., Matthews, J. T., McNamara, R. S. and McPhewn, M. P. (1994) 'Feeding 10 billion people in 2050: The key role of the CGIAR's International Agricultural Research Centers', *A Report by the Action Group on Food Security*, CGIAR, Washington, DC

Zanago, R. G. (2003) 'Farmers see no risk in "Golden Rice"', *Manilla Bulletin*, vol VII, p32

Zandstra, H. and Taylor, P. (2006) 'Farming systems research: A retrospective consideration of the importance of ecological scale and stakeholder participation', CRDI, Ottawa, Canada

Zimmermann, M. J. D. O. (1996) '"Commentary"' on R. D. Graham and R.M. Welch (1995) '"Breeding for staple food crops with high micronutrient density"', in H. E. Bouis (ed.) *Working Papers on Agricultural Strategies for Micronutrients No.3*, International Food Policy Research Institute, Washington, DC

Zimmermann, R. and Qaim, M. (2004) 'Potential health benefits of Golden Rice: A Philippine case study', *Food Policy*, vol 29, pp147–68

Index

Acharya, A.K. 36
actor-networks 7, 9–10, 14, 93, 109, 116, 137
adaptive research 3, 5
ADB (Asian Development Bank) 9, 12, 43, 54, 94, 111, 114, 127, 132
Adelaide University (Australia) 38, 43, 45, 49
Adorada, Dante 57
African Rice Centre (WARDA) 62, 109
agriculture-nutrition-health 1–2, 12, 32–40, 110–111, 117, 125, 131
    CGIAR and 15, 38–40
    micronutrient approaches 36–40, 50
    phases of 34–36
Al-Babili, S. 75, 85, 86–87, 89, 122
aleurone layer 60–61
anaemia 36, 37, 48, 51, 54
    effects of 71–72
Anderson, R.S. 2, 6, 16–19, 22–24, 41, 47, 66, 115, 136, 139, 141
anthropology 7–8
anti-nutrients 47, 131
Apthorpe, R. 4, 7
Aquino, Corazon 26
Arabidopsis 6
ARIs (advanced research institutes) 97, 99
aromatic rice varieties 48–49, 59, 60
Ashby, J.A. 31, 41
Asia 19, 83, 84, 86, 101
    in Cold War 16–17
    scientists in 70–71
Asian Rice Biotechnology Network 70
Australia 45, 70

Bacillus thuringiensis (Bt) corn *see* Bt maize
banana 101, 104–105, 121, 122
Bangladesh 54, 85
barkada 57–58
Barker, R. 63
barley 6, 101
Barry, Gerard 85, 91, 108, 109, 120
Bayer Crop Science 65
beans 47, 100, 101, 107, 119
Beard, John 54, 57
Beijing Genomics Institute 6

Bellagio conferences (1969–71) 21
'best bets' 138, 139
beta-carotene 68, 71, 76–77, 85, 86, 90, 92, 121, 122, 136
Beyer, Peter 74–75, 76, 78, 80, 81, 82, 83, 85, 86–87, 91, 94, 122
Biggs, S.D. 23–24, 25
bioavailability 45, 47, 49, 50, 52, 54, 62, 67, 86, 89, 92, 113, 122
bioefficacy studies 54–59, 112, 121
biopolitics 4–5, 17, 44, 52, 53, 62–67, 119, 136
biosafety 88
black boxing 3, 10, 12, 44, 62, 66, 74, 93, 108, 109, 116, 121, 133, 137
BMGF (Bill and Melinda Gates Foundation) 4, 9, 41, 92, 138, 140
    'Grand Challenges' 9, 104–105, 121–122
    and HarvestPlus 39–40, 94, 99–105, 121–122, 130
    investment strategy of 101–102, 103–104
    and NGOs 101
Borlaug, Norman 19
Bouis, Howarth 12, 34, 35, 38, 98, 100, 109, 114, 120
    and HarvestPlus 81, 98, 100, 105, 109, 114, 120
    and iron rice 53, 54, 57, 58, 61, 94, 135
    on micronutrients 43, 44, 45, 46, 47–48, 138
boundaries 7, 8, 12, 14, 81, 87, 111, 112, 124, 135
Brader, L. 29
Brady, Nyle 22
Brazil 98, 120
brown rice 111, 112, 134
Bruntland Report (1987) 25
Bryce, J. 33
Bt corn *see* Bt maize
Bt maize 88–89, 90
Buffett, Warren 102–103

C4 variety 57, 60
Cakmak, I. 45
Calloway, D.H. 117, 132
calories 44, 51

Cantrell, R.P. 6, 28
cassava 47, 101, 104–105, 122
Castillo, G.T. 52, 63–64, 90
Catholic convents/congregations 53–59
centres of excellence 29–30, 31, 115, 116, 124
CGIAR (Consultative Group on International
  Agricultural Research) 2, 3, 6, 9, 15, 75, 125,
  126, 130
    broker/gatekeeper role 4, 97–98, 114, 115,
      124, 137
    bureaucratization of 22
    centres of excellence in 29–30, 32, 115, 124
    Challenge Programs see Challenge Programs
    directors-general 18, 22, 24–25, 26, 28, 29,
      32
    eco-regional approach 25
    evolution of 12, 16–17, 21
    'farmer first' movement/FSR and 23, 24–26
    funding activities of 21, 25, 27, 44, 69
    and Golden Rice 81
    and iron rice 54, 66
    'mega-programmes' of 138–139
    micronutrients project (1994–9) 13, 34,
      38–40, 43, 44–45, 49, 94, 111, 117, 142
    and NARS 25, 26–27
    reform of 4, 21, 27–28, 94–99, 124, 131
    Science Council see Science Council
    site specificity and 25, 30–31
    SW/EPs in 28–29
    system-wide initiatives in 28–29, 95, 96
    TAC and 21, 22, 25, 27, 94–95
    see also CIMMYT; IRRI
Challenge Programs (CGIAR) 3, 7, 9, 13, 28,
  29–32, 131
    Biofortification see HarvestPlus
    funding for 29, 32, 38, 41
    and micronutrients 38
    origin/definition of 95–97
    top-down/centralized approach of 31, 41,
      116
Chambers, R. 15, 24
Chandler, Robert 18, 21, 26
Chataway, J. 29, 32
chemical fortification 52, 53
  see also food fortification
child malnutrition 32, 33, 42, 51, 72, 76, 118
    and education 39, 53
    and income 35
China 9, 64, 69, 83, 85, 89, 120
    HarvestPlus and 98, 120–121
CIAT (International Center for Tropical
  Agriculture) 21, 47, 94, 99, 100, 105
CIMMYT (International Maize and Wheat
  Improvement Center) 16, 21, 47, 69
    QPM 34, 38, 44–45
'classic cluster' 18, 23, 41, 134
Clay, E.J. 23–24, 25
Cocodrie variety 85, 86, 90
Cohen, J. 36
Cold War 16–20
Combs, G.F. 12, 45
comparative advantage 3

consumer choice see 'end-user' question
cooking/cooked rice 56, 62
Copenhagen Consensus (2004) 39, 72, 131
corn see maize
Cornell University (New York) 9, 38, 43, 45, 81,
  82, 126
cost-benefit analysis 36, 71
cost-effectiveness 1, 36, 38–39, 72, 99, 119, 131
CPs see Challenge Programs
Cribb, J. 48, 132
cropping systems approach 21–24
Cullather, N. 17, 18, 19, 20, 87
Cuu Long Delta Rice Research Institute 83, 84,
  86, 122

daffodil gene 85
DALY (disability-adjusted life year) 36, 118–119,
  131
DANIDA (Danish International Development
  Agency) 48
Danish Environmental Assessment Institute 39
Datta, Swapan 75, 84, 85, 107, 108
Dawe, D. 63, 64
'definitive centre' 6, 17–18, 22, 24, 32, 69
de Kadt, E. 35
Delisle, D. 5
Del Mundo, Angelita 53–54, 57, 58, 107, 126
Derham, Robert 76
diet 37, 46–47, 51, 52, 62, 76, 121, 142
direct-seeded rice 5
disease 26, 36, 39, 49, 51, 60, 71
    see also anaemia; GBDS; VAD
diversity in farming 5, 23
donor agencies 9, 21, 26, 28
Dubock, Adrian 76, 79, 80, 81, 91
DuPont 26

ecological diversity 5
economies of scale 37, 72
eco-regional concept 25, 28–29
Edwards, M. 2
Eicher, C.K. 27–28, 32, 41, 124
'Ending Hidden Hunger' conference (Ottawa,
  1991) 43, 52, 72, 73
'end-user' question 14, 106–107, 117–118, 119,
  131, 132, 140–141
energy gap 33, 34
epistemology 7, 24
ethnography 7–8
ETH (Swiss Federal Institute of Technology) 9,
  13, 68, 69, 71, 75–76, 78, 108–109
Europe/European Union (EU) 70, 75

Fan, Yunliu 120
FAO (Food and Agriculture Organization) 21, 37
'farmer first' movement 24–25, 31
farmer-scientist partnerships 27, 31, 41, 140
farmer-to-farmer training 25
farming, small-scale/subsistence 1, 20–21, 65,
  117, 131
'fast-tracked' varieties 107, 108
FCND (Food Consumption and Nutrition

Division, IFPRI) 44
Felix, Angelina 53, 57, 58
Ferguson, J. 8
Fernandez, Doreen 90
ferritin (Ft) 56, 113
Fiedler, J.L. 73
fields, ethnographic 8
fields, research 10–12
Fitzhugh, H. 29
'fixed genetic potential' view 37, 39
   see also malnutrition
Flood-Prone Rice Research Programme (IRRI) 48
Florencio, C.A. 51, 73–74
FNRI (Food and Nutrition Research Institute, Philippines) 51, 52, 53
food crisis 1–2, 16, 18, 33, 36, 138
food fortification 5, 37, 51–52, 101
   and carbonated drinks 51
   and cooking oil 51, 52, 73
   and margarine 51, 73
   and MSG (monosodium glutamate) 73
Food Fortification Act (Philippines, 2000) 52–53, 63, 66
'food, health and care' framework (UNICEF) 35, 36
food prices 34, 52, 64, 65
'Food for School' programme (Philippines) 53, 66
food security/insecurity 3, 33, 34, 63–66, 67, 95
food systems approach 12, 45–47
Ford Foundation 6, 16, 17, 21, 75, 126
fortification of foods see food fortification
Foucault, M. 5
framing 4–5, 7, 10, 11–12, 87–91, 93, 140
'freedom to operate' 78–80, 81
Freiberg University (Germany) 74, 85, 122
FSR (farming systems research) 23, 24
Fujisaka, S. 25
Future Harvest Centers 97–98, 99, 106

Gasper, D. 4
Gates Foundation see BMGF
Gaud, William S. 18
GBDS (Global Burden of Disease Study) 36
genetics see plant genetics
Germany 74, 85, 122
germplasm banks 99
Gillespie, S. 36, 71–72
Glaeser, B. 19
Global Burden of Disease (World Bank) 33
   see also GBDS
global health 33, 35–36, 39
'global' research field 7–8, 9, 13–14, 32, 125, 126–127, 140–141
Gloria Rice 64–65
glutinous rice 59
GM (genetic modification) 69–71, 76, 101, 133–135
   and iron rice 109, 112–113
   Golden Rice see Golden Rice
   opposition to 6, 13, 68, 77, 87, 88–89, 92
Golden Rice 6, 7, 28, 68–92
   appearance of 68

beta-carotene levels of 85, 86, 89
bioavailability of 89, 92
controversies surrounding 76–83
efficacy of 76–77, 78, 81, 89
ETH involvement in 9, 13, 68, 69, 71, 74–75
farmers' acceptance of 89–91, 92
'freedom to operate' review of 78–80, 81
funding for 13, 68, 75
and GM debate 68
and HarvestPlus 14, 81, 94, 107, 108–109, 117, 119, 122, 124, 126, 137
Humanitarian Board see Humanitarian Board
humanitarian licence for 69, 79, 84
IP/TP of 77, 79, 80
IRRI and 69, 78, 79, 82, 83, 84, 85–86, 87–88, 89
in Philippines 85, 87–91, 132
post-harvest stability of 13, 89, 90
as 'proof of concept' study 81–83, 91, 136
prototype 75–76, 84, 85–86
and public-private partnership 73, 76, 83, 91
release of 83–91
research teams/network for 70–71, 75, 81, 84–85
Rockefeller Foundation/IPRB and 8, 13, 68, 69–70, 71, 78–79
Syngenta/Zeneca and 68, 76, 79, 80, 81, 86, 87
taste/texture factors of 90
and technology transfer 77–78, 82, 91
governance issues 2, 8, 29, 91, 105, 123, 137
Graham, Robin 45, 46–47, 49, 57, 61, 120, 135
grain mineral content 50, 56, 60
   see also iron rice/high-iron rice; IR68144 germplasm
'Grand Challenges' (BMGF) 9, 104–105
'Green Evolution' 26
Greenland, D.G. 25
Greenpeace 78, 89
Green Revolution 1–2, 3, 12, 18–21, 63, 65, 69, 139
   agriculture-nutrition linkage in 34
Gregorio, Glen 49, 51, 56–57, 60, 62, 107, 111, 114
Griffin, K. 19
Gupta, A. 8
GxE (genotype by environment) 12, 61–63, 67, 107, 111, 113, 116, 134

Haas, J.D. 53, 54, 55, 56, 57, 62
Hall, A. 32
HarvestPlus 1, 5, 15, 41–42, 67, 93–124, 130, 133, 136–137, 138
   breeding targets 109, 110–111, 114
   and China 98, 120–121
   'end-user' question in 14, 106–107, 117–118, 119
   evolution/naming of 94–99, 105–107
   funding for 9, 39–40, 41, 94, 99, 99–105, 106, 107, 115, 120, 136
   and Golden Rice project see under Golden Rice

governance/oversight mechanism for 105, 123
and IPGs 94, 95, 97, 116
and iron rice project *see under* iron rice
and IRRI 107–109, 114
partnerships in 93–94, 99, 106, 107,
109–114, 115, 122–123, 126, 128
policy/impact issues in 116–119
research strategies of 6, 7, 109
significance of 40
sweet potato initiatives 107
tensions in 107, 110, 114–116, 124, 137,
142
Hawkes, C. 40
Hb 56
healthcare, access to 35
Herdt, Robert 71, 75
Hervé, Philippe 108–109
high-iron rice *see* iron rice/high-iron rice
high-lysine maize 34, 44
see also QPM
Hilchey, T. 48
HKI (Helen Keller International) 73
Hossain, Mahubub 69
HRCP (Hybrid Rice Commercialization Program)
64–65
Humanitarian Board (Golden Rice project)
80–81, 83, 85, 86, 87, 91, 94, 118, 122–123,
129
humanitarian licence 69, 79, 84
hunger 16, 18, 33, 65–66
Hunt, Joseph 54
hybrid rice 64–65, 67, 128
see also HRCP
subsidies for 64, 65–66
Hybrid Rice Commercialization Program *see*
HRCP
HyRice Corp 65
HYVs (high yielding varieties) 1–2, 19, 23, 26

IARCs (International Agricultural Research
Centres) 27
see also CGIAR
IDD (iodine deficiency disorders) 71
IFPRI (International Food Policy Research
Institute) 9, 29, 43, 94, 98, 105, 126
agriculture-nutrition-health studies 34, 38, 40
FCND *see* FCND
and micronutrients 45–46
IFR (iron-fortified rice) 52–53, 66
IHNF (Institute of Human Nutrition and Food,
UPLB, Philippines) 43, 53, 57, 58
IITA (International Institute of Tropical
Agriculture) 21
immunization 72
see also NIDS
impact 4, 5, 9, 14, 39, 41, 67, 95, 107, 117–121
and context 130–133
and spin-offs 119–121
import of grains 19, 53, 64, 66
'Improving Human Nutrition through Agriculture'
seminar (1999) 38, 49, 54, 94
India 19, 48, 49, 75, 83, 85, 98, 120

Green Revolution in 24–25
semi-dwarf varieties in 20
indica varieties 85, 87, 89
'individual adaptability' view 37
see also malnutrition
Indonesia 54, 64, 83, 85
Indo-Swiss Collaboration on Biotechnology 83
industrial food fortification 37
see also food fortification
industrialization 19
innovation system 29-30
intellectual property *see* IP
interdisciplinary division 62, 66, 67, 141
interdisciplinary integration 9, 14, 41, 44, 67,
70–71, 93–94, 124, 128–130, 134, 139, 141
international crop research 12, 15, 18, 94, 116,
126–127
centres, founding of 21
FSR in 24
and MDGs 3, 4
neutrality of 23
partnerships 4, 40, 114, 125–128, 140–142
as a public good *see* public goods
see also CGIAR
international development 39, 125
research for 29, 95
science/technology and 2–3, 12
targets *see* MDGs
International Program on Rice Biotechnology *see*
IPRB
International Year of Rice 5, 63, 67
Investing in Health (World Bank, 1993) 36
iodine 36, 71–72
iodine deficiency 71–72
IPGs (international public goods) 3–4, 13, 14,
29–31, 41, 47, 83–84, 94, 95, 97, 116,
117–118, 125, 130, 135–136, 140
IP (intellectual property) rights 77, 79, 80, 84,
108, 112
IPRB (International Program on Rice
Biotechnology) 9, 68, 69–70, 75, 76, 126, 127
IR8 variety 19–20, 87
IR64 variety 85, 86, 90
IR72 variety 48, 49, 50
IR68144 germplasm 48–50, 53–63, 107–108,
110–111
agronomic characteristics of 49–50, 63
and cultural practices 56, 57
bioefficacy study/'feeding trial' 53–59
certification/release of 59–60, 63
contradictory views of 60–63
and environmental and seasonal factors 56
and Golden Rice 85
grain iron content/distribution 60–61
and post-harvest practices 56, 60
yield 50, 59, 60, 66
iron deficiency anaemia *see* anaemia
iron rice/high-iron rice 1, 8, 9, 12, 14, 28, 43–67,
94, 126–127
bioavailability of 45, 47, 49, 50, 52, 54, 63, 67
and Catholic convents/congregations 43,
53–59, 66

and cultural practices 56, 57, 61
funding of 12, 43, 54
grain iron content/distribution 60–61
GxE factors in 61–63, 67
HarvestPlus and 62, 67, 100–101, 107–108, 109, 121, 122
and interdisciplinary inquiry 44, 66, 67, 128
and politics 63–67
and post-harvest practices 56, 60, 61, 112
and proof of concept 43, 44
'research family' 43, 57–59, 66, 81, 94, 108–109, 111, 135
transgenic approach to 109, 112–113
see also IR68144 germplasm
irrigation 18, 48
IRRI (International Rice Research Institute) 9, 15, 17–18, 69, 127
broker/gatekeeper roles 127
bureaucratization of 22
cropping system approach of 21–24
'farmer first' movement and 24–25
founding of 12, 16, 126
funding for 6, 12, 21, 22, 28
and Golden Rice see under Golden Rice
and Green Revolution 19–21
and HarvestPlus 107–109, 114
and host country (Philippines) 6, 8, 13, 25–26
HYVs developed by 19
and iron rice research 49, 54, 55–59, 66, 107–109, 112–114, 116
laboratories 112
and micronutrients project (1994–9) 38, 47, 48–49
mission-oriented research in 22
and NGOs 88
restructuring of 25–26
and rice blast scandal 26
rice-naming policy 23
salinity tolerance breeding programme 43
staff cuts in 28
and plot-lab methodological model see plot-lab methodological model
ISAAA (International Service for the Acquisition of Agri-biotech Applications) 78–79, 80
IVACG (International Vitamin A Consultative Group) 74

Japan 28, 64, 70
japonica variety 84, 85, 89
Jenny, Katharina 81
Johnson-Welch, C. 40

Kaybonnet variety 85, 86
Kennedy, E. 35
Khush, G.S. 48
King, John 17
Knight, J. 133
Knorr-Cetina, K. 14
Korea 64
Kryder, R.D. 79

Lampe, Klaus 25, 26

Lancet 32, 33, 41
Latin America 2, 19, 25, 101
Latour, B. 3, 9, 12, 137
Lei, Xingen 120
linear/non-linear models 4–5, 13, 44, 67, 77, 80, 87, 91, 129, 132
Lipton, M. 35
local factors 7, 8, 23, 31, 131–132, 140
see also impact; site specificity
Lopez, A.D. 36
Los Baños (Philippines) 6, 8, 10, 53, 126
see also UPLB
lysine 34, 44, 122
see also high-lysine maize; QPM

Macapagal-Arroyo, Gloria 64–65
McNamara, Robert 21
maize 6, 34, 47, 101, 120
gene, in Golden Rice 85
GM 88, 90
see also Bt maize
see also high-lysine maize; QPM
Malaysia 64
malnutrition 5, 36, 45
child see child malnutrition
'fixed genetic potential' view of 37, 39
and income 35, 40
'individual adaptability' view of 37
measurement of 36
Manila (Philippines) 43, 53, 54
MAP (Mexican Agricultural Program) 16, 17, 18
Marcos, Ferdinand 20, 26, 65, 87–88
Marcus, G.E. 8, 10, 11
Mason, J.B. 72, 73
maternal health 33, 35, 42, 51
MDGs (Millennium Development Goals) 3, 4, 33, 34, 38, 41, 66, 72, 99, 130, 131, 136, 138
media 20, 25–26, 66, 76, 108
medical model 5
'mega-programmes' 138–139
see also CGIAR
Mestizo variety 65
see also hybrid rice; HRCP; 'Gloria Rice'
Mexico 3, 69
micronutrients 36–42, 47
and CGIAR 38–40, 43
deficiencies, effects of 36, 45
food systems approach to 45–46
funding for projects 48
interventions 5, 13, 33, 34, 36, 43–48
and public health 71–72
and soil see soil and micronutrients
see also iron rice/high-iron rice; Golden Rice, VAD, anaemia
mineral toxicity 48
modernization of agriculture 16, 19
Monsanto 7, 80, 85, 108
mortality rates 33, 71, 72
MS13 variety 60, 63, 65–66, 107–108, 111, 132
see also iron rice/high-iron rice; IR68144 germplasm
MS (Maligaya Special) varieties 59

MTAs (material transfer agreements) 79
Murray, C.J.L. 36

Narciso, Melanie 53, 57
NARES (national agricultural research and
    extension systems) 97, 99
NARS (National Agricultural Research Systems)
    25, 26–27, 31, 41, 48, 127, 139
    and Golden Rice 78, 79, 84, 85
    and HarvestPlus 114–115
    in Philippines see PhilRice
NFA (National Food Authority, Philippines) 52, 53
national health systems 35
national security/sovereignty 16, 17, 20, 90
Nestle, Marion 76–77, 81, 82
networks 8, 9–10, 14, 94, 114, 121, 126, 141
NGOs (non-governmental organizations) 7, 9, 26,
    36, 64, 73, 79, 99
    and CGIAR 98
    and IRRI 88
    sustainable development 88–89
NIDs (national immunization days) 72
Normile, D. 70–71
NRM (natural resource management) 27, 32, 96
NSIC (National Seed Industry Council) varieties
    59, 60
Nueva Ecija (Philippines) 26
nutrition 5, 32–42, 62, 76–77, 126
    competing paradigms for 33, 37
    as 'global' problem 39, 122
    goal-oriented approach to 36, 37
    and income 35
    investments in 38–39, 41
    Lancet on 32, 33, 41
    micronutrients approach see micronutrients
    and plant breeders 41, 62, 110, 134
    retention, post-harvest 12, 13
    rights-based approach to 37
    and time allocation 35, 40
    and women/children 32, 33, 35, 42
    and women's empowerment 35, 40
    and yield 34, 44–48
    see also agriculture-nutrition-health
Nutrition Act (Philippines, 1974) 51
NVASP (National Vitamin A Supplementation
    Program, Philippines) 73

Oasa, E.K. 23, 24
Oryza sativa see rice
O'Toole, J.T. 70

Pacey, A. 37
Padolina, W.G. 60
patents 79, 80
Payne, P. 37
Perkins, J.H. 16, 20
philanthropists 4, 38, 39, 102–104
Philippines 9, 17, 19, 49
    biosafety framework in 88
    certification in 59–60, 132
    diet in 47, 51–52, 62, 142
    FNRI see FNRI

Food Fortification Act (2004) 52–53, 66
GM maize 88–89, 90
    see also bt maize
Golden Rice in 85, 87–91, 132
hybrid rice 64–66
    see also HRCP
iron rice project in see iron rice
and IRRI 6, 8, 13, 87–88, 126
NFA (National Food Authority) 52–53
Nutrition Act (1974) 51
rice crisis in 63–64
Rice Enrichment Law (1952) 51
rice imports/self-sufficiency in 53, 63–64, 65,
    66, 67
rice politics in 20, 25–26, 63–67
Science and Technology Dept./Agriculture
    Dept. (DOST/DA) 88, 89
varietal testing programme (RVIG/NCT) in
    59
vitamin A supplement program in see NVASP
PhilRice (Philippine Rice Research Institute) 9, 26,
    43, 64–65, 70–71, 89, 115–116, 122
Pinstrup-Andersen, P. 29, 30–31, 34
plant breeding/breeders 6, 23, 34, 38, 53,
    113–114, 133
    in micronutrient projects 44–45, 47, 48
    and nutritionists 41, 62, 110, 134
    targets 109, 110–111, 114
plant genetics 2, 5, 6, 18
    and micronutrients 47
plant nutrition 45, 46, 47
plot-lab methodological model 22
PNSL (Plant Nutrition and Soil Laboratory) 45
Poletti, Susanna 109
polished rice 60, 111, 112, 134
population growth 16, 17
Population-National Security Theory 16
positioning of research 10–12
post-harvest practices 12, 13, 56, 60, 61, 112,
    118, 135
Potrykus, Ingo 74–76, 77, 78, 80–81, 86, 87, 94,
    108
poverty 1–2, 16
    and malnutrition 33, 39, 117–118
    reduction strategies 96, 97, 125
power-knowledge relations 8, 24, 62, 91, 140
Pretty, J.N. 15, 24
private sector 2, 7, 65, 69, 133
    see also public-private partnerships
productivity 2, 3, 65
'proof of concept' 6, 13, 43, 44, 53–59, 81–83,
    91, 107–108, 114, 121, 122, 130, 134, 136, 138
pro-poor strategies 13, 90, 98, 118, 119, 123
protein era 33, 34, 44, 122
pro-vitamin A 1, 28, 68, 71, 74, 76, 82, 85
PSBRc28 variety 56–57, 60
PSB Rc 82/128 varieties 90
PSNL (Plant, Soil and Nutrition Laboratory) 38,
    47, 126
public goods 3–4, 13, 14, 29
    international see IPGs
public health

and agriculture 12, 28, 32, 35
    environmental 35
    global 33, 35
    and structural adjustment 35
    and vitamin A deficiency 71–74
    see also agriculture-nutrition-health
public-private partnerships 2, 7, 29, 51, 73, 76, 83, 91, 97–98, 101, 114, 127, 135
PVMRC (ProVitaMinRice Consortium) 9, 115, 122–123, 126, 133

Qaim, Matin 118, 133
QPM (Quality Protein Maize) 34, 38, 44–45
    see also high-lysine maize
QRs (quantitative restrictions) 64

Ramos, Fidel 52, 73
Randazzo, Fil 138
Reeves, T.G. 6
rice biotechnology 6, 13, 69–70, 84–85
    see also IPRB
rice blast 26
rice crisis (Philippines, 1995) 63–64
Rice Enrichment Law (Philippines, 1952) 51
Rice, International Year of 5, 63, 67, 86
'rice is life' slogan 17
rice markets 5–6, 17, 52, 53, 64, 66, 117
rice milling/millers 51, 52, 56, 57, 60, 67, 111, 112, 134
rice-naming policy 23
rice (*Oryza sativa*) 5–7
    genome sequences 6
    importance of 5, 6, 17, 51
rice prices 52, 64, 65
rice varieties 5, 9, 13
rights-based approach 37
Rockefeller Foundation 6, 9, 16, 21, 102, 126, 133
    definition of agriculture 17
    and Golden Rice project 13, 68, 69–70, 71, 75, 78-79, 83
    and rice biotechnology *see* IPRB
    role in establishing the CGIAR/IRRI 16–17, 21
Rodin, Judith 102
Rogers, E.M. 80
Rothschild, George 28
Ruel, M.T. 40
Rukuni, M. 27–28, 32, 41, 124
RVIG (Rice Varietal Improvement Group, Philippines) 59

Sachs, Jeffrey 39, 102
saffron rice 90
Saith, A. 3, 39
Salas, Rafael 20
salt iodization programmes 33, 36, 51, 71, 72
salt tolerance breeding programme 43, 48
Sangkap Pinoy Seal Program 52
scale 7, 119
    'scaling up' 33, 107, 132–133, 138, 139
Schaffer, B. 7, 31, 82, 136, 139
Science Council (CGIAR) 29, 30, 39–40, 41, 94–95, 97, 107, 133

science policy processes 7–12
    and actor-networks 7, 9–10, 14
    and anthropology/ethnography 7–8
    and knowledge production 7, 8, 14
    and locations/positioning 10–12
    and negotiated boundaries 7, 8, 12, 14
    upstream–downstream relations in 12, 14
scientific excellence 29–30, 32, 115, 124
    see also centres of excellence; 'definitive centre'
Sebastien, Leocardo 70–71
seed certification 19, 59–60
seed dormancy/vigour 50
self-sufficiency 17, 19, 20, 63–64, 65, 67
semi-dwarf varieties 19–20
Sen, Amartya 34
Senadhira, Dharmawansa 48–49, 57, 126
Serageldin, Ismail 25, 26
Shore, C. 8
Simmonds, N.W. 134
Sison, Cristina 57
site specificity 7, 8, 23, 25, 27, 30–31, 37, 135, 140
SL Agritech 65
small-scale/subsistence farming 1, 20–21, 65, 117, 131
soil and micronutrients 45, 46, 48, 50, 56, 61–62
Sommer, A. 71
sorghum 101, 104–105, 122
South/Southeast Asia 2, 10, 17, 19, 64, 90, 129
sovereignty 20, 90
Spitz, P. 18
Sri Lanka 19, 48, 64
Stansfield, Sally 100
Stein, A.J. 117, 118–119
Stonesifer, Patty 103–104
Strathern, M. 7
stress tolerance 66, 70
structural adjustment 26, 35
Sub-Saharan Africa 2, 27
subsidies 64, 65–66
sugar 51, 52, 73
Sumberg, J. 115
Swaminathan, M.S. 24–25
sweet potato 1, 101, 107, 114, 119, 132
SW/EPs (system-wide/ecoregional programmes) 25, 28–32, 38, 94
Swiss Federal Institute of Technology *see* ETH
Syngenta 6, 7, 9, 68, 76, 79, 80, 81, 86, 87, 127
    and CGIAR 98
'system-wide' initiatives *see* SW/EPs

TAC (Technical Advisory Committee) 21, 22, 25, 27, 29, 38, 94–95
taste/texture factors 13, 19, 90
    see also Golden Rice
Taylor, P. 24, 25
technicism 19
technology-first approach 17–18
technology transfer 70, 77–81, 82, 91
Time magazine 76, 108
Toenniessen, Gary 75, 80, 81
Tolentino, V.B.J. 64

TP (technical property) rights 77, 79, 80
transgenic methods *see* GM
Truman, Harry S. 16
Tsing, A. 7, 9
tungro disease 49, 60
Turkey 45

uncertainty 5, 9, 12, 13, 41, 44, 46, 66, 67, 87, 116
  see also black boxing
UNDP (United Nations Development Programme)
  21
UNICEF 35, 36, 71
United Nations (UN) 3, 5, 36, 39
  Millennium Project *see* MDGs
United States (US) 19, 70, 85, 89
  foreign policy 16–17, 18, 20
  and MAP 16, 17
UPLB (University of the Philippines, Los Baños)
  43, 53, 126
USAID (US Agency for International
  Development) 16, 18, 38, 81, 119

VAD (vitamin A deficiency) 36, 71–74, 76, 92
  effects of 72, 77
Van Roozendaal, G. 38
Vietnam 53, 54, 64, 83, 84, 85, 122
Virk, Parminder 109
VITAA (Vitamin A for Africa) 107, 119
vitamin A
  absorption of 76–77
  bioavailability of 74
  deficiency *see* VAD
  foods fortified with 1, 51, 52
  injections/supplements of 77
  policy uncertainties and 73–74
  pro- *see* pro-vitamin A
  programmes 33, 46, 72, 100–101
  as public health problem 71–74

and rice biotechnology *see* Golden Rice
  taste/texture factors 90
vitamin B 51, 52
vitamin E 92, 122
Voss, Joachim 100

Waite Institute (Adelaide University) 38, 45, 48,
  126
WARDA (African Rice Centre) 62
washing rice 56, 57
Water and Food Challenge Program 97
Welch, Ross 45, 46–47, 57, 61, 120, 135
wheat 6, 47, 101, 119, 120
  HYVs of 2, 19
  iron-enriched 51, 52
  vitamin A-enriched 52, 73
  zinc-enriched 1, 45, 107
WHO (World Health Organization) 35, 100
women and nutrition/health 33, 35, 40, 42, 48,
  51, 118
World Bank 21, 33, 34, 35, 41–42, 81, 82, 99
World Development Report 36
World Food Day 86
Wright, S. 8

Ye, Xudong 75
yield 17, 23, 56, 66, 136
  and certification 59, 60
  and nutrition 34, 44–48
  see also HYVs

Zandstra, Hubert 24, 25
Zawa Bonday variety 48, 49
Zeneca *see* Syngenta
Zimmermann, M.J.D.O. 38
Zimmermann, R. 133
zinc 43, 46–47, 49, 50, 55, 92, 100–101
zinc-efficient wheat 1, 45, 107